Lecture Notes in Computer Science 1220

Edited by G. Goos, J. Hartmanis and J. van Leeuwen

Advisory Board: W. Brauer   D. Gries   J. Stoer

Springer
Berlin
Heidelberg
New York
Barcelona
Budapest
Hong Kong
London
Milan
Paris
Santa Clara
Singapore
Tokyo

Peter Brezany

Input/Output Intensive Massively Parallel Computing

Language Support, Automatic Parallelization, Advanced Optimization, and Runtime Systems

 Springer

Series Editors

Gerhard Goos, Karlsruhe University, Germany

Juris Hartmanis, Cornell University, NY, USA

Jan van Leeuwen, Utrecht University, The Netherlands

Author

Peter Brezany
Institute for Software Technology and Parallel Systems
University of Vienna
Liechtensteinstrasse 22, A-1090 Vienna, Austria
E-mail: brezany@par.univie.ac.at

Cataloging-in-Publication data applied for

Die Deutsche Bibliothek - CIP-Einheitsaufnahme

Brezany, Peter:
Input output intensive massively parallel computing : language
support, automatic parallelization, advanced optimization, and runtime
sytems / Peter Brezany. - Berlin ; Heidelberg ; New York ; Barcelona
; Budapest ; Hong Kong ; London ; Milan ; Paris ; Santa Clara ;
Singapore ; Tokyo : Springer, 1997
  (Lecture notes in computer science ; Vol. 1220)
  ISBN 3-540-62840-1 kart.

CR Subject Classification (1991): D.3, D.1.3, D.4.2, C.1.2,F.1.2, B.3.2,
D.4.4, G.1

ISSN 0302-9743
ISBN 3-540-62840-1 Springer-Verlag Berlin Heidelberg New York

Typesetting: Camera-ready by author
SPIN 10549551   06/3142 – 5 4 3 2 1 0   Printed on acid-free paper

*To my wife Jarmila, children Zuzana and Jozef, my
mother Zuzana, and in memory of my father Jozef*

Preface

Massively parallel processing is currently the most promising answer to the quest for increased computer performance. This has resulted in the development of new programming languages, and programming environments and has significantly contributed to the design and production of powerful massively parallel supercomputers that are currently based mostly on the distributed-memory architecture.

Traditionally, developments in high-performance computing have been motivated by applications in which the need for high computational power clearly dominated the requirements put on the input/output performance. However, the most significant forces driving the development of high-performance computing are emerging applications that require a supercomputer to be able to process large amounts of data in sophisticated ways. Hardware vendors responded to these new application requirements by developing highly parallel massive storage systems.

However, after a decade of intensive study, the effective exploitation of massive parallelism in computation and input/output is still a very difficult problem. Most of the difficulties seem to lie in programming existing and emerging complex applications for execution on a parallel machine.

The efficiency of concurrent programs depends critically on the proper utilization of specific architectural features of the underlying hardware, which makes automatic support of the program development process highly desirable. Work in the field of programming environments for supercomputers spans several areas, including the design of new programming languages and the development of runtime systems that support execution of parallel codes and supercompilers that transform codes written in a sequential programming language into equivalent parallel codes that can be efficiently executed on a target machine. The focus of this book is just in these areas; it concentrates on the automatic parallelization of numerical programs for large-scale input/output intensive scientific and engineering applications. The principles and methods that are presented in the book are oriented towards typical distributed-memory architectures and their input/output systems.

The book addresses primarily researchers and system developers working in the field of programming environments for parallel computers. The book will also be of great value to advanced application programmers wishing to gain insight into the state of the art in parallel programming.

Since Fortran plays a dominant role in the world of high-performance programming of scientific and engineering applications, it has been chosen as the basis for the presentation of the material in the text.

For full understanding of the contents of the book it is necessary that the reader has a working knowledge of Fortran or a similar procedural high-level programming language and a basic knowledge of machine architectures and compilers for sequential machines.

The book's development In writing this book, I utilized the results of my work achieved during research and development in the European Strategic Program for Research and Development in Information Technology (ESPRIT), in particular, ESPRIT Projects GENESIS, PPPE, and PREPARE. Most of the methods and techniques presented in the book have been successfully verified by a prototype or product implementation or are being applied in on-going projects. Topics related to parallel input/output have been the basis for the proposals of new research projects that start at the University of Vienna this year.

The material of the book has been covered in courses at the University of Vienna given to students of computer science and in the Advanced Course on Languages, Compilers, and Programming Environments given to advanced developers of parallel software.

Contents of the book Each chapter begins with an overview of the material covered and introduces its main topics with the aim of providing an overview of the subject matter. The concluding section typically points out problems and alternative approaches, discusses related work, and gives references. This scheme is not applied if a chapter includes extensive sections. In this case, each section is concluded by a discussion of related work. Some sections present experimental results from template codes taken from real applications, to demonstrate the efficiency of the techniques presented.

Chapter 1 provides motivation, a brief survey of the state of the art in programming distributed-memory systems, and lists the main topics addressed in the book. Input/Output requirements of the current Great Challenge applications are illustrated in three examples which are both I/O and computational intensive: earthquake ground motion modeling, analysis of data collected by the Magellan spacecraft, and modeling atmosphere and oceans.

Chapter 2 specifies a new parallel machine model that reflects the technology trends underlying current massively parallel architectures. Using this machine model, the chapter further classifies the main models used for programming distributed-memory systems and discusses the programming style associated with each model.

The core of the book consists of chapters 3–7. While the first chapter deals with programming language support, the subsequent three chapters show how programs are actually transformed into parallel form and specify requirements on the runtime system. Chapter 7 develops new concepts for an advanced runtime support for massively parallel I/O operations.

Chapter 3 describes Vienna Fortran 90, a high-performance data-parallel language that provides advanced support both for distributed computing and the operations on files stored in massively parallel storage systems. In this chapter the presentation is mainly focused on the language extensions concerning parallel I/O.

Chapter 4 first describes the principal tasks of automatic parallelization of regular and irregular in-core programs and then addresses several important optimization issues. In-core programs are able to store all their data in main memory.

Chapter 5 deals with basic compilation and optimizations of explicit parallel I/O operations inserted into the program by the application programmer.

Chapter 6 treats the problem of transforming regular and irregular out-of-core Vienna Fortran 90 programs into out-of-core message-passing form. Out-of-core programs operate on significantly more data (large data arrays) that can be held in main memory. Hence, parts of data need to be swapped to disks.

Compilation principles and methods are presented in chapters 5 and 6 in the context of the *VFCS (Vienna Fortran Compilation System)*.

Chapter 7 proposes an advanced runtime system referred to as *VIPIOS (VIenna Parallel Input/Output System)* which is based on concepts developed in data engineering technology.

Chapter 8 (conclusion) presents some ideas about the future development of programming environments for parallel computer systems.

Acknowledgements Many people deserve thanks for their assistance, encouragement, and advice. In particular, I thank Hans Zima for giving me the opportunity to work on the ambitious research and development projects and for his encouragement. I am grateful to Michael Gerndt (KFA Jülich), Erich Schikuta, and Thomas Mück (both from the Department for Data Engineering of the University of Vienna), Joel Saltz (University of Maryland), and Alok Choudhary (Northwestern University, Evanston) for many stimulating discussions and suggestions, and to Viera Sipkova for the implementation of several parallelization methods introduced in this book within the Vienna Fortran Compilation System. Thanks also go to Minh Dang and Ka Heng Wong for carrying out many performance measurements to demonstrate the efficiency of the techniques presented in the book.

It is also a pleasure to thank colleagues from GMD Bonn, GMD FIRST Berlin, GMD Karlsruhe, Parsytec Aachen, TNO Delft, ACE Amsterdam, TU Munich, INRIA Rennes, Steria Paris, University of Liverpool, and University of Southhampton with whom I have cooperated in the ESPRIT projects GENESIS, PREPARE, and PPPE.

January 1997   *Peter Brezany*

Table of Contents

1. Introduction

High Performance Computing is currently one of the leading edge technologies in the information age. In computer systems, massive parallelism is the key to achieving high performance. Consequently, massively parallel processing and its use in scientific and industrial applications has become a major topic in computer science. This has added a new impetus to computer architectures and has resulted in the design of new programming languages, new algorithms for expressing parallel computing and in the development of programming environments supporting efficient production of parallel software.

The first means of introducing massive parallelism into the design of a computer has been the approach taken in the construction of *SIMD* (Single Instruction Stream, Multiple Data Streams) computers, like MPP Goodyear, Connection Machine and MasPar. SIMD computers work best when performing large-scale array operations; they are not suitable for general purpose computation.

The dominant architecture in the 1990s is *distributed-memory MIMD* (Multiple Instruction Streams, Multiple Data Streams) *systems* (DMSs). These systems comprise hundreds of *nodes* including *processors* and *memory*, and high-speed and scalable *interconnects*. Currently, a number of extremely powerful supercomputers of the DMS type are commercially available. They offer a peak performance up to several hundred GFlops and a main memory capacity of several hundred Gbytes. Teraflops systems with terabytes of memory are expected in the coming years.

Recent developments show that high performance computing systems take many different forms, including DMSs, *symmetric shared-memory machines (SMPs)*, workstation clusters, and networks of SMPs with hybrid parallelism. Furthermore, the improvement of global communication links has led to the possibility of computing in geographically distant *heterogeneous distributed networks*.

Supercomputers are commonly interfaced with various peripheral devices for pre- and postprocessing of data or simply for additional working storage. In many cases, the speed of access to data in an external storage system can determine the rate at which the supercomputer can complete an assigned job. Disk drives are the major components of an I/O subsystem. Over the past decade, significant advances in disk storage device performance have been achieved. Inexpensive, high-capacity disks are now making it possible to construct large *disk arrays* with high peak performance. The bandwidth of I/O operations increased dramatically by performing them in parallel. Technology advances, furthermore, include increasing the density of storage media such as disks and tapes. The mass storage industry has developed technologies for handling petabytes (10^{15} bytes) of data today and is developing technologies for handling hexabytes (10^{18} bytes) in the future.

Currently, massively parallel systems are almost indispensable to solve large scale research problems in physics, chemistry, biology, engineering, medicine and various other scientific disciplines. However, it is still necessary to further extend the application area of high performance computing. In the past the high cost of high performance systems has been a barrier for their wider usage, but the availability of low-cost commodity microprocessors is driving design costs down and, in the words of Richard M. Karp from the University of California at Berkeley, it may not be long before high-performance parallel computing becomes widely available on the desktop. Consequently, this development will further increase the acceptance of massively parallel systems and contribute to extension of the range of applications of high performance parallel computing.

The fundamental challenge is to make massively parallel systems widely available and easy to use for a wide range of applications. This is crucial for accelerating the transition of massively parallel systems into fully operational environments. A significant amount of software research for achieving these objectives is currently underway both in academia as well as industry. The research effort can be broadly categorized into four classes, namely languages, parallelizing compilers, runtime systems, and support tools.

All of these research areas, except support tools are addressed in this book. The realization that the parallel I/O problems have to be addressed with the same determination and creativity as the parallel computation problems has been our motivation for our effort, the results of which are presented.

1.1 I/O Requirements of Parallel Scientific Applications

With the increase in CPU speeds and inter-processor communication bandwidth, more data can be processed per unit time, requiring more data to be transferred between a node and

- secondary storage,
- network interfaces,
- other nodes, and
- special devices, e.g., visualization systems.

Consequently, high performance computing systems handle substantially more data – both input and output – than traditional systems. Large-scale simulations, experiments, and observational projects generate large multidimensional datasets on meshes of space and time. Accurate modeling requires that the mesh be as dense as possible. These applications require that relevant subsets of information in large datasets be accessed quickly – in seconds or minutes rather than hours. Therefore, providing high computing power and inter-processor communication bandwidth together with large quantities of main memory *without* adequate I/O capabilities would be insufficient to tackle a multitude of additional real world problems, called *I/O intensive* applications[1]. As a matter of fact, most I/O subsystems of existing DMS are hardly able to cope with today's performance requirements. Moreover, even the most elaborated state-of-the-art mass storage systems of today as attached to leading edge massively parallel machines will quickly run out of steam when facing the forthcoming supercomputing challenges involving terabytes of raw data, i.e., satellite data or streams of seismic data emerging from a multitude of sensor devices. As a result, many scientific applications have become I/O bound on the current generation of highly parallel computing platforms, i.e. their run-times are dominated by the time spent performing I/O operations. Consequently, the performance of I/O operations has become critical for high performance in these applications.

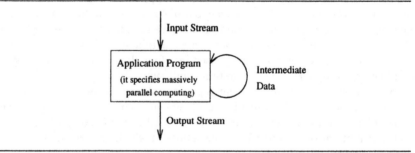

Fig. 1.1. I/O Intensive Application from an Application Developer's Perspective

Many computer architects are now focusing their attention on the performance problems of I/O. Over the past decade, significant advances in disk storage device performance have been achieved. The need for high performance I/O is so significant that almost all present generation parallel computing systems such as the Paragon, iPSC/860, Touchstone Delta, CM-5, SP-1, nCUBE2, Parsytec GC, Meiko CS-2 etc. provide some kind of hardware support for parallel I/O.

Fig. 1.1 depicts the main components that a typical I/O intensive and computation intensive application includes from an application developer's

---

[1] In general, an application is referred to as *I/O intensive* if it has substantial I/O requirements [153].

perspective: application program specifying the massively parallel computing, input data stream, movement of intermediate data generated during program execution, and the output stream.

From the published work in this area [100, 117, 218, 230] and our own analysis of I/O intensive applications, it is expected that most of the cases when I/O operations should be inserted by the application developer and/or parallelizing compiler can be classified into one of the following categories:

o *I/O is inherent in the algorithm*
  - *Input*
    Most programs need to read some data to initialize a computation. The input data varies in size from small files containing a few important parameters to large data files that initialize key arrays. An application may require input data only at initiation of the computation or, alternatively, at intervals throughout the computation.
  - *Output*
    Output from a program takes many forms; it can be a small file with a few results or a large file containing all the values from several arrays. Like input, output may occur only at the completion of program execution or, alternatively, periodically at intervals throughout the computation when time-dependent data are involved. Output must often be generated for use on sequential computer systems, e.g., comparison of earlier results or use in workstation-based visualization systems.

o *Checkpoint/Restart*
  For long-running production codes, it is desirable to have the ability to save the state of the computation to a file or several files in order to continue computing from that point at a later date. In the special case of checkpoint/restart, a key question is whether the restart will occur with the same number of processors as the checkpoint.

o *Memory Constraints*
  - *Scratch Files*
    Scratch files are often used to hold data structures containing intermediate calculations that a user does not wish to recompute and that do not fit into main memory.
  - *Accessing Out-of-Core Structures*
    An application may deal with large quantities of primary data which cannot be held in the main memories of computational nodes. These data structures (e.g., large arrays) must be swapped to disks and are called *out-of-core structures*. They are brought into the main memories in parts, processed there, and parts of results are transferred to disks. On some machines, there is hardware and software support for virtual memory; the performance, however, achieved is often very poor.

In many cases, the speed of access to data determines the rate at which the supercomputer can complete an assigned job. Today, most high-performance applications involve I/O rates of 1 to 40 Mbytes per second (MB/sec) for secondary storage and 0.5 to 6 MB/s for archival storage. Application developers indicate that probably 1 GB/sec to secondary storage and 100 MB/sec to archival store will be required in the near future. To better understand the need for such high data-transfer rates, we provide the following three examples.

1.1.1 Earthquake Ground Motion Modeling

The reduction of the earthquake hazard to the general population is a major problem facing many countries. To this end, it is essential that within earthquake-prone regions new facilities be designed to resist earthquakes and existing structures be retrofitted as necessary. Assessing the free-field ground motion to which a structure will be exposed during its lifetime is a critical first step in the design process. It is now generally recognized that a complete quantitative understanding of strong ground motion in large basins requires a simultaneous consideration of three-dimensional effects of earthquake source, propagation path, and local site conditions. These needs place enormous demands on computational resources, and render this problem among the "Grand Challenges" in high performance computing.

Bao et al. [18] describe the design and discuss the performance of a parallel elastic wave propagation simulator that is being used to model earthquake-induced ground motion in large sedimentary basins. They report about the using the code to study the earthquake-induced dynamics of the San Fernando Valley in Southern California. The San Fernando simulations involve meshes of up to 77 million tetrahedra and 40 million equations. The code requires nearly 16 GB of memory and takes 6 hours to execute 40,000 timeyear coupled steps on 256 processors of the Cray T3D. I/O is used at the beginning of the program to input the files storing the mesh data and every tenth time step to output results to disk. In the current program implementation, I/O is serial.

1.1.2 Image Analysis of Planetary Data

This application develops a parallel program which performs image analysis of data collected by the Magellan spacecraft. Magellan was carried into Earth orbit in May 1989 by the space shuttle Atlantis. Released from the shuttle's cargo bay, it was propelled by a booster engine toward Venus, where it arrived in August 1990. Magellan used a sophisticated imaging radar to pierce the cloud cover enshrouding the planet Venus and map its surface. During its 243-day primary mission, referred to as Cycle 1, the spacecraft mapped over 80 percent of the planet with its high-resolution Synthetic Aperture Radar. The spacecraft returned more digital data in the first cycle than all previous

U.S. planetary missions combined. By the time of its third 243-day period mapping the planet in September 1992, Magellan had captured detailed maps of 98 percent of the planet's surface.

Magellan has transmitted in excess of 3 Tbytes of data to earth. Producing a 3D surface rendering at 200 Mbytes of data per frame would require more than 13 Gbytes per second at 50 frames per second. This exceeds the I/O capacity of today's machines by far. Rendering a portion of the Venusian surface on a 512-node Intel Touchstone Delta takes several days [136, 153].

1.1.3 Climate Prediction

Current atmosphere/ocean models place specific demands on supercomputers [153]. Consider one such application, *Climate Prediction using General Circulation Model (GCM)*. For a 100-year atmosphere run with 300-square kilometers resolution and 0.2 simulated year per machine hour, the simulation on an Intel Touchstone Delta takes three weeks runtime and generates 1.144 Gbytes of data at 38 Mbytes per simulation minute. For a 1,000-year coupled atmosphere-ocean run with a 150-square kilometer resolution, the atmospheric simulation takes about 30 weeks and the oceanic simulation 27 weeks. The process produces 40 Mbytes of data per simulation minute, or a total of 20 Tbytes of data for the entire simulation.

1.1.4 Summary

Table 1.1 summarizes an estimate of the current requirements of a few large-scale computations[2]. These requirements are expected to increase by several orders of magnitude in most cases.

The interested reader can find an annotated bibliography of papers and other sources of information about scientific applications using parallel I/O in [171].

1.2 State-of-the-Art and Statement of the Problems

Performance of any high-performance system includes hardware and software aspects.

At the software level, there is a need for languages and programming environments that will allow application programmers to realize a significant fraction of machines' peak computational and I/O performance. By now, DMSs have traditionally been programmed using a standard sequential programming language (Fortran or C), augmented with message passing constructs. In this paradigm, the user is forced not only to deal with all aspects of the distribution of data and work to the processors but also to control the

---

[2] For more details please refer to [100] and references provided there.

Table 1.1. Examples Illustrating Current I/O Requirements of Some Grand Challenge Applications (A: Archival Storage, T: Temporary Working Storage, S: Secondary Storage, B: I/O Bandwidth)

Application	I/O requirements
Environmental modeling	T: 10s of GB. S: 100s of MB - 1 GB per PE. A: Order of 1 TB.
Eulerian air-quality modeling	S: Current 1 GB/model, 100 GB/application; projected 1 TB/application. A: 10 TB at 100 model runs/application.
Earth system model	S: 108 MB/simulated day. 100 GB/decade-long simulation. B: 100 MB/s.
4-D data assimilation	S: 100 MB - 1 GB/run. A: 3 TB database. Expected to increase by orders of magnitude with the Earth Observing System (EOS) - 1 TB/day.
Ocean-Atmosphere Climate modeling	S: 100 GB/run (current). B: 100 MB/sec. A: 100s of TB.
Solar activity and heliospheric dynamics	B: 200 MB/sec A: Up to 500 GB
Convective turbulence in astrophysics	S: 5-10 GB/run B: 10-100 MB/s
Particle algorithms in cosmology and astrophysics	S: 1-10 GB/file; 10-100 files/run. B: 20-200 MB/s
Radio synthesis imaging	S: 1-10 GB A: 1 TB

program's execution by explicitly inserting message passing operations. The resulting programming style can be compared to assembly language programming for a sequential machine; it has led to slow software development cycles and high costs for software production. Moreover, although MPI (Message Passing Interface) is evolving as a standard for message passing, the portability of MPI-based programs is limited since the characteristics of the target architecture may require extensive restructuring of the code.

For efficient handling parallel I/O[3], vendors usually use a *parallel file system* approach that aims at increased performance by declustering data across a disk array thereby distributing the access workload over multiple servers. Generally it can be said that parallel file systems lack the notion of a data type and give only limited possibilities to influence the data declustering scheme. Another way considered for providing support for parallel I/O are *high-level parallel I/O libraries*. These libraries are often built on top of parallel file systems. Many existing runtime I/O systems for parallel computers have tended to concentrate on providing access to the file system at a fairly low level; users often need to know details of disk-blocking and read-write caching strategies to get acceptable performance. Recently, an MPI-IO initiative started, whose aim is to choose the appropriate extensions to MPI to

---

[3] From the software point of view, parallel I/O means providing access by a collection of processes to external devices.

support the parallel I/O portable programming. This approach is based on the idea that I/O can be modeled as message passing: writing to a file is like sending a message, and reading from a file is like receiving a message.

The enormous importance of DMSs and the difficulty of programming them has led to a considerable amount of research in several related areas both in academia as well as in industry. A long-term goal of that research is to provide programming environments at least comparable in power, comfort and generality to that available at present day workstations.

Various efforts have been made to implement a virtual shared memory on top of a DMS and this way make programming DMSs as easy as programming shared memory systems. This can be done either in hardware or by appropriate software mechanisms.

NUMA systems offering a shared address space on top of a distributed-memory hardware architecture have appeared on the market, and there is a trend towards hybrid architectures - clusters of shared-memory processors, which integrate shared-memory parallelism at the node level with distributed-memory parallelism. Language extensions such as SVM Fortran [32] and Fortran-S [34] allow the user to provide assertions towards the exploitation of locality and the minimization of data transfers.

A dominating method for providing a virtual shared memory for a DMS is automatic parallelization: in this approach, the user is enabled to write a code using global data references, as on a shared memory machine, but she or he is required to specify the distribution of the program's data. This data distribution is subsequently used to guide the process of restructuring the code into an SPMD (Single–Program– Multiple–Data) [156] program for execution on the target DMS. The compiler analyzes the source code, translating global data references as stated in the source program into local and non-local data references based on the data distributions specified by the user. The non-local references are resolved by inserting appropriate message-passing statements into the generated code.

Fortran is being used as the primary language for the development of scientific software for DMSs. On the other hand, an important problem faced by application programmers is the conversion of the large body of existing scientific Fortran code into a form suitable for parallel processing on DMSs. In 1992, a consortium of researchers from industry, laboratories funded by national governments and academia formed the *High Performance Fortran Forum* to develop a standard set of extensions for Fortran 90 which would provide a portable interface to a wide variety of parallel architectures. In 1993, the forum produced a proposal for a language called *High Performance Fortran (HPF)* [110, 168], which focuses mainly on issues of distributing data across the memories of a DMS. The main concepts in HPF have been derived from a number of predecessor languages, including DINO [225], CM Fortran [92], Kali [167], Fortran D [116], and *Vienna Fortran (VF)* [273], with the last two languages having the strongest impact. Typically, these languages allow the explicit specification of *processor arrays* to define the set of (abstract) processors used to execute a program. *Distributions* map data arrays to processor sets; the establishment of an *alignment* relation between arrays results in the mapping of corresponding array elements to the same processor, thus avoiding communication if such elements are jointly used in a computation.

Within the past few years, a standard technique for compiling HPF languages for distributed memory has evolved. Several prototype systems have been developed, including the *Vienna Fortran Compilation System (VFCS)* [74], which was among the very first tools of this kind.

In HPF-1, the current version of HPF, there is sufficient support for computationally intensive regular grid-based applications via regular data distribution and alignment annotations, thus alleviating the task of the programmer for a sizable segment of applications.

A significant weakness of HPF-1 and HPF-1 programming environments is their lack of support for applications, which are I/O intensive or are irregular. Many large-scale scientific applications, characterized by features like multiblock codes, unstructured meshes, adaptive grids, or sparse matrix computations, cannot be expressed in HPF-1 without incurring significant overheads with respect to memory or execution time.

From the computer-science point of view, an application is considered to be *irregular* if there is limited compile time knowledge about data access patterns and/or data distributions. In such applications, major data arrays are accessed indirectly. This means that the data arrays are indexed through the values in other arrays, which are called indirection arrays. The use of indirect indexing causes the data access patterns, i.e. the indices of the data arrays being accessed, to be highly irregular, leading to difficulties in determining work distribution and communication requirements. Also, more complex distributions are required to efficiently execute irregular codes on DMSs. Researchers have developed a variety of methods to obtain data mappings that find data distributions which enable optimization of work distribution and communication. These methods produce irregular distributions. Irregular codes can be found in the number of industrial and research applications that range from unstructured multigrid computation fluid dynamic solvers, through molecular dynamics codes and diagonal or polynomial preconditioned iterative linear solvers, to time dependent flame modeling codes. The task of language designers and compiler developers is to find an elegant way for specification and efficient methods for implementation of irregular applications.

Current research is going into an extension of HPF supporting irregular computations, parallel I/O and task parallelism. Vienna Fortran and Fortran D already provide support for irregular computations that includes irregular data distribution schemes and parallel loops with work distribution specification. One of the distribution schemes uses *indirect distribution func-*

tions that allow the specification of a general distribution via a mapping array. Moreover, Vienna Fortran provides a higher level language support and VFCS provides an interactive interface which enable the user to direct the appropriate programming environment to derive irregular data and work distributions automatically according to a selected partitioning strategy. The research goal is to develop the appropriate compilation methods capable of handling real application codes. Basic compilation methods for irregular applications were already implemented in both VFCS and the Fortran D compiler [57, 64, 62, 256].

This fact has been acknowledged by the HPF Forum in its decision to start the development of HPF-2 [111] at the beginning of 1995, and has motivated a range of other research projects. In October 1996, the forum produced a proposal for a new HPF language specification [143].

Parallel I/O support primitives have been included into Vienna Fortran [55] and later, similar constructs were proposed for HPF [39, 236]. However, so far they have not been approved by any implementation. Recently, a draft proposal for language support for out-of-core structures was elaborated [45, 244]. Compilation techniques proposed so far for out-of-core programs only focus on very simple code patterns.

Checkpointing of codes developed for massively parallel systems presents a challenging problem of great importance because processing these codes very often tends to fail [213, 233, 234]. *Checkpointing* is the act of saving the volatile state of a program (e.g. registers, memory, file status) to stable storage (e.g. disk) so that after a failure, the program may be restored to the state of its most recent checkpoint.

The major use of checkpointing is to allow long-running programs to be resumed after a crash without having to restart at the beginning of the computation. Another application is *parallel program debugging* (see below). By checkpointing at moderate intervals, like once an hour, a long-running program is guaranteed to lose a maximum of one hour of running time to each system failure. This is preferable to restarting the program from scratch at each failure.

Checkpointing and resuming can be provided to the application programmer in two modes:

1. *Providing a transparent recovery scheme*
   The checkpoint algorithm runs concurrently with the target program, interrupts the target program for fixed amounts of time and requires no changes to the target's code or its compiler. Upon a system failure, the programmer needs only to restart the program with the appropriate command line.

2. *Providing a semi-transparent recovery scheme*
   Providing a transparent recovery scheme is an attractive idea and some
   efficient schemes can be found in the literature. However, a semi-transparent
   method may present some important advantages, namely
   - it optimizes the storage needed for saving checkpoints since it just saves
     the data that is really essential for recovery in case of failure.
   - as a consequence, it reduces the performance overhead induced by
     checkpoints.
   - since the programmers are responsible for the placement of checkpoints
     they can tune the fault-tolerance according to their needs.
   - a semi-transparent scheme can be more easily portable to other sys-
     tems.
   The disadvantage of a semi-transparent approach is requiring some effort
   from the application programmer - she or he has to annotate the program
   with checkpointing directives.

A long-term research topic is the development of *compiler assisted techniques*
[181] for checkpointing.

Another application for a real-time concurrent checkpoint techniques is
parallel program debugging. Multiple checkpoint of an execution of the target
program can be used to provide "playback" for debugging; an efficient, real-
time checkpoint technique can provide this playback with minimal impact on
the target program execution.

So far, no attempt has been made to control checkpoint/restart in HPF-
like languages.

HPF-1 as well as Vienna Fortran are based upon the data parallel SPMD
paradigm. Many applications require a more general model of computation,
which often needs to be combined with the data parallel model (e.g., multidisciplinary optimization specifies a number of tasks, each of which is a
data parallel program). While task management has been a topic of research
for several decades, particularly in the operating systems research commu-
nity, there has not been much attention given to the mechanisms required for
managing control parallel tasks, which may themselves be data parallel.

Research in this direction has been performed in many projects, including
Orca [16],Fortran-M [114], Fx [242, 243], OPUS [71, 139].

1.3 The Central Problem Areas Addressed in this Book

This book considers automatic parallelization of I/O intensive applications
for massively parallel systems as an integrated process which includes handling parallel I/O aspects and parallel computation aspects. To deal with
automatic parallelization, one needs first of all to consider the following three
components:

1. language
2. compilation system that includes a parallelizing subsystem
3. runtime support

which are the main concerns of this book.

The basis for the synergy of parallel I/O and parallel compute features is a suitable parallel machine model which the language design, the development of compilation techniques and the design of an advanced runtime system is based on. In the book, we present a parallel machine model that integrates the current trends in the massively parallel architecture development. The realization of the designed language framework is illustrated by the Vienna Fortran language extensions. However, the approach can be applied to any language that is based on the same programming model. Further, the appropriate compilation techniques developed are introduced in the context of the Vienna Fortran Compilation System.

In the presentation of compilation techniques, a strong focus is given to irregular applications. The compilation techniques are divided into three classes:

1. compiling in-core programs
2. compiling parallel I/O operations
3. compiling out-of-core programs

whose common features are identified and addressed in the design of the optimization transformations.

The runtime system support issues arise when developing compilation techniques of each class. In the book, a separate chapter is devoted to the runtime system support for parallel I/O operations and out-of-core programs.

In this book we do not deal with checkpoint/restart techniques

2. Parallel Programming Models

Modeling sequential or parallel computation includes two components: a machine model and a programming model[1]. A *machine model* defines the structure and behaviour of an execution engine in sufficient detail that it is possible to determine which among alternative solutions is best, i.e. it allows performance estimates. A *programming model* provides abstractions for formulating agorithms that extend the capabilities of an underlying machine model. The abstractions raise the conceptual level of programming while incurring "known" costs. A programming model is implemented by an appropriate programming language.

2.1 The Parallel Machine Model

An appropriate machine model, or type architecture [237], plays an irreplaceable role in the programming language design, specification of compilation methods and application software development. The machine model abstracts away the irrelevant details of physical computer systems while retaining the appropriate truth to be able to give a rough idea of the execution speed of a program. The abstract machine that is represented by the model need not be the isomorphic image of the physical machine. Consequently, the specific object architecture need not be of concern and it can change.

Sequential scientific computing relies on the *von Neumann machine model* that is an efficient bridge between software and hardware. Several attempts (e.g., [93, 184, 203, 239, 250, 238, 253, 253, 254]) were made to design analogous models for parallel computing.

In this section, we specify a new parallel machine model called *IMPA* (mnemonic for *I*ntegrated *M*assively *P*arallel *A*rchitecture), that reflects the technology trends underlying massively parallel architectures including parallel I/O subsystems. Further, we examine briefly how closely our model corresponds to reality on current parallel computer systems. Because parallel computers are much more disparate than sequential machines, the "distances" between the machine model and some systems will be greater than

---

[1] It can be seen from the literature that the terms machine model and programming model mean diferent things to different people. In this book we use similar vocabulary as in [239, 238]

one expects in the sequential case. Especially, I/O subsystems of different massively parallel systems are architecturally very different. Nevertheless, in this section, we specify a general model that includes parallel I/O. It will serve as a basis for the classification of the existing programming models (Subsection 2.2.3), the design of new language constructs (Chapter 3), the specification of the compilation methods (Chapters 4, 5 and 6), and the design of the runtime support (Chapter 7).

2.1.1 IMPA Model

At the highest level of abstraction, we postulate IMPA as a machine including a massively parallel computational part and a massively parallel I/O part (see Fig. 2.1).

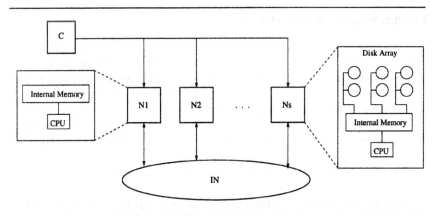

Fig. 2.1. The IMPA Machine Model

Definition 2.1.1. *The IMPA is a system S = (N, C, IN, T), where*

- N *is the set of* nodes *that perform parallel computational and parallel I/O tasks.*
  Each node includes a processor *(CPU) and* hierarchical memory, *called the* local memory. *It consists of internal* [2] *(core) memory and, in some cases, external* [3] *(out-of-core) memory (see Fig. 2.2). The processors work asynchronously. According to the memory hierarchy and functionality, the nodes are classified into 3 classes (levels):*
  1. Pure computational nodes *(subset* U *of* N*). They include a processor and internal memory only.*

---

[2] This is very often called the main or primary memory.
[3] This is very often called the disk or peripheral memory.

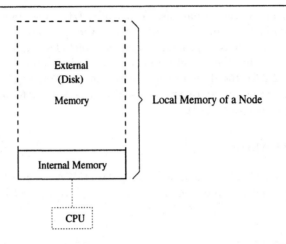

Fig. 2.2. Structure of Local Memory

2. **Pure I/O nodes** *(subset* W *of* N*). They include a processor, internal memory, and a set of disks (two-level memory model [4]). However, the processor is responsible for controlling I/O activities only. Therefore, it is often referred to as the* I/O *processor.*

3. **Combined nodes** *(subset* W *of* N*). They include a processor, primary memory, and a set of disks. The processor fulfills both computational and I/O tasks.*
   Obviously, the following equation holds: $N = U \cup V \cup W$

- IN *is the* **interconnection network** *of unspecified topology that connects the nodes. The connections establish* **channels** *over which* **messages** *are sent between nodes. A channel is an abstraction of a physical communication network; it provides a communication path between processors. Channels are accessed by means of two kinds of primitives: send and receive.*

- C *is the* **controller** *that is linked to the nodes of* N *by a common, narrow channel. Its purpose is to model controllers (hosts) of the real machines and to encapsulate certain global capabilities that should be provided in hardware (barrier synchronization, broadcast, parallel prefix, etc.).*

- T *is the set of parameters characterizing interprocessor communication and the access time to secondary storage. Each parameter characterizes the cost of a data reference relatively to the local primary memory fetch cost. It is assumed that CPUs reference their local primary memories in unit time,*

---

[4] In general, it would be possible to consider an n-level memory model as in [253, 254].

$t_0 = 1$. *The IMPA model further stipulate that a processor's reference to data that is not in its own primary memory requires a latency[5] of t_1 time units to be filled if the data is currently in primary memory of a remote processor, t_2 time units if the data is in own disk memory and t_3 time units if the data is in disk memory of a remote processor. It is assumed that*

$$t_3 \gg t_2 \gg t_1 \gg 1.$$

The T parameters may be determined for any concrete architecture.

For example, for the Intel Paragon System, the following access time limits have been measured:

local memory > 120 nanoseconds
remote memory (through message passing) > 120 microseconds
(shared) disk arrays > 23 milliseconds

The IMPA features force programmers and algorithm designers to attend to the matter of

- locality and data reusing, i.e. how to arrange computations so that references to data in local internal memory are maximized, and the time-consuming references to data on non-local disks are minimized;
- hiding communication and I/O latency by computations.

2.1.2 Matching the IMPA to Real Architectures

In the last subsection, we introduced an idealized machine model. Now, we very briefly discuss how faithfully the IMPA models contemporary DMSs. We will see that every feature of the IMPA has been implemented either directly or can be simulated reasonably cheaply.

First, we discuss a typical uniprocessor architecture, showing the interconnection between processors and I/O devices. Then, we provide a review of disk arrays that represent a fundamental form of parallel I/O; Chen et al. [76] and Gibson [121] provide more detailed surveys. Finally, we give an overview of architectural features of existing DMSs[6].

A typical uniprocessor architecture

A computer system consists of various subsystems which must have interfaces to one another. For example, the memory and processor need to communicate, as do the processor and the I/O devices. This is commonly done with a *bus*. A bus is a shared communication link, which uses one set of wires to

---

[5] Latency is the delay between the request for information and the time the information is supplied to the requester.

[6] An excellent introduction to multiprocessor I/O architectures is given by Kotz in [173].

connect multiple subsystems. The two major advantages of the bus organization are versatility and low cost. By defining a single connection scheme, new devices can easily be added, and peripherals can even be moved between computer systems that use the same kind of bus.

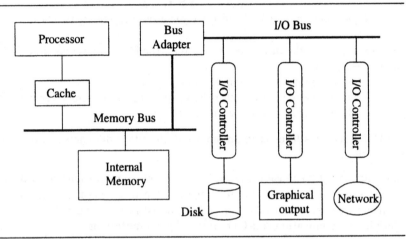

Fig. 2.3. Interconnection between the Processor and I/O Devices

Fig. 2.3 shows the structure of a typical uniprocessor system with its I/O devices (a more detailed depiction and explanation can be found in [209], for example). The *bus adapter* bridges between the proprietary CPU–memory bus and an I/O bus. The I/O bus is typically based on a standard such as *Small Computer System Interface (SCSI)*[7]. A *cache* is, in general, faster memory used to enhance the performance of a slower memory taking advantage of locality of access. *Controllers* connect the I/O bus to specific I/O devices (disk, network, or graphics). The controllers are responsible for the low-level management of the device, interpreting standard I/O commands from the bus. The disk controller mediates access to the disk mechanism, runs the track-following system, transfer data between the disk and its client, and, in many cases, manages an embedded cache. All of the components from the individual I/O devices to the processor to the system software affect the performance of tasks that include I/O.

Disk arrays

One promising direction for future disk systems is to use arrays of smaller disks that provide more bandwidth through parallel access. *I/O bandwidth* can be measured in two different ways ([209]):

---

[7] SCSI arose from the cooperation of computer manufacturers.

1. How much data can be moved through the system in a certain time?
2. How many I/O operations can be done per unit time?

Which measurement is best may depend on the environment; in scientific applications, transfer bandwidth is the important characteristic.

To improve the capacity and bandwidth of the disk subsystem, we may group several disks into a *disk array*, and distribute a file's data across all the disks in the group. This practice is typically called *striping, declustering*, or *interleaving*. There is no universal agreement on the definition of these terms, but common usage, according to [173], seems to indicate that *declustering* means any distribution of a file's data across multiple disks, whereas *striping* is a declustering based on a round-robin assignment of data units to disks. *Interleaving* is less commonly used now, but some have used it to mean striping when disks are rotationally synchronized.

Disk arrays were proposed in the 1980s as a way to use parallelism between multiple disks to improve aggregate I/O performance.

Early work of Kim [159] and Salem [227] demonstrated the usefulness of disk arrays, but one of the significant drawbacks was reduced reliability.

In 1988 Patterson, Gibson, and Katz presented "a case for redundant arrays of inexpensive disks (RAID)" [208], in which they argued that disk arrays could be faster, cheaper, smaller, and more reliable than traditional large disks, and categorized several techniques for using redundancy to boost the availability of disk arrays.

Today disk arrays appear in the product lines of most major computer manufacturers.

Example architectures

Below, we introduce an overview of architectural features of several existing DMSs[8], from the past and present. All the systems discussed have some sort of either one or a set of controllers that are either directly linked to the nodes, or the communication between the controllers and the nodes takes part through the network.

The Meiko CS-2 System. ([187]). All the IMPA node types may occur in the CS-2 configuration. Currently, all CS-2 systems have 2 independent network layers, each a complete, independent, data network. The architecture supports up to eight layers.

The Intel iPSC/2 and Intel iPSC/860 Systems. ([14, 48, 212, 219]). The iPSC/2 System with I/O consists of a hypercube-connected set of computational nodes with an attached set of I/O nodes and a host computer. All

---

[8] All of them, but the Intel Touchstone Delta System, are commercial systems.

nodes are interconnected by the Direct Connect Network so that any node can communicate directly with any other node. I/O nodes do not run application processes directly, but provide disk services for all users. Each disk I/O node has 4 Mbytes of memory and a SCSI bus which connects the I/O processor to one or more high capacity disk drives. The disk drives have an internal multiple-block buffer.

The Intel Touchstone Delta System. The Intel Touchstone Delta architecture is a 16 x 32 two-dimensional array of computational nodes with 16 I/O nodes on each side of the array; each I/O node has 2 disks associated with it.

The Intel Paragon System ([90]). The Intel Paragon, on the other hand, has a mesh topology. Separate I/O nodes are used which can be placed anywhere within the mesh network. Each I/O node has its own set of disks attached to it as in our model; the set is maintained as a RAID level 3 device [76].

The NCUBE System ([219]). The NCUBE architecture is based on a separate board, called a "NChannel Board", to support the I/O subsystem. A NChannel board contains a ring of 16 I/O nodes, using the same processors as used for computational nodes. Each of the I/O nodes may control a single 4 MB/sec channel to a disk controller (or other I/O device controller) which handles 1-4 disks. Each of the 16 I/O nodes is also connected to 1–8 computational nodes.

The CM-5 System ([144, 177, 180]). The CM-5 architecture is based on a set of pure computational nodes and a set of pure I/O nodes. The nodes are connected by a fat-tree (see [176], page 33). Computational nodes are grouped into *partitions*, and each partition is assigned a special processor node working as a *partition manager*. The I/O nodes are specialized for different kinds of I/O services; there are "disk" nodes, "tape" nodes, etc. Each disk storage node has a controller (based on a SPARC CPU), a CM-5 network interface, a buffer RAM, and four SCSI bus adapters, typically with two disks each.

Parsytec Power GC ([107, 178]). The system consists of a computational kernel and a set of I/O nodes connected by an I/O network. The computational kernel consists of computational nodes connected by a data network. The I/O network connects all computational nodes with I/O nodes.

The IBM SP2 System ([6, 126]). SP2 nodes can be configured to have one of two fundamental logical personalities. Some nodes are configured as *computational nodes* and are used for executing user jobs. Other nodes are configured as *server nodes* that provide various services required to support the execution of user jobs on computational nodes; these could be integrated file servers, gateway servers for external connectivity, database servers, backup and archival servers, etc. There are parallel file system server I/O nodes and standard file server I/O nodes. External servers are connected to the SP2 via intermediate gateway nodes.

2.2 Programming Models for Distributed-Memory Systems

A machine model, which in principle could be physically implemented, typically describes the computation at a too low level for convenient programming and algorithm design. A programming model in general extends a machine model with an additional set of abstractions that raise the conceptual level of programming. The benefits of abstraction mechanisms to programmers need not involve substantial performance costs. A programming model is implemented by an appropriate programming language. The language should assist the programmer in the abstraction process by confining computer dependencies to specific objects only. Moreover, the language facilities implementing the abstractions have "known" costs in the sense that programmers can have a rough idea how expensive the facilities are to use, and that the costs are reasonably independent of which physical machine runs the program.

A *sequential programming model* extends the von Neuman machine model. It determines which data types are available to the programmer, how to structure a program into program units, how data flow between these units, how to determine the control flow of a program, etc. Its abstraction mechanisms include the ability to declare variables, which are abstractions of state data, and procedures, which are new state operators and functions.

Specifically, a *parallel programming model* should determine additional features (Giloi [123]): which art of parallelism should be utilized (task or data parallelism), which program elements (program parts) should be executed in parallel[9], how the data is laid out in internal and external memory, how to distribute the work, how to organize communication between program parts that are executed in parallel, how to coordinate the parallel run.

Programming models for DMSs can be classified according to the role of the user in the program development process. Exploiting the full potential of DMSs requires a cooperative effort between the user and the language system. There is a clear trade-off between the amount of information the user has to provide and the amount of effort the compiler has to expend to generate optimal code. At one end of the spectrum are message passing languages where the user has to provide all the details while the compiler effort is minimal. At the other end of the spectrum are sequential languages where the compiler has the full responsibility for extracting the parallelism.

---

[9] We often talk about the *granularity* in this context.

For the past few years, median solutions, such as Vienna Fortran, Fortran D and HPF have been explored, which provide a fairly high level environment for DMSs while giving the user some control over the placement of data and computation.

There are essentially two parallel models that are being used in programming scientific applications for DMSs, the *message-passing programming model*, and the *data parallel programming model*. They both are discussed in a greater detail in parts 2.2.1 and 2.2.2, respectively. In some cases, the message passing approach can also be applied to parallel I/O. However, these models are certainly not the only approaches that can be taken for representing parallel computation. Many other models have been proposed ([113, 122]), differing in the abstraction level they provide to the programmers. One of them is a programming model providing coordination primitives which is briefly introduced in part 2.2.3.

2.2.1 Message-Passing Programming Model

The implementationally simplest and computationally most efficient programming model for distributed-memory architectures is the message-passing model, for this reflects directly the features of the machine model. Many different language mechanisms have been proposed for specifying parallel execution, communication, and synchronization when implementing this programming model in high-level languages.

In the message-passing paradigm, parallel execution is performed at the level of communicating processes[10] [147] that have their own private address space and share data via explicit messages; the source process explicitly sends a message and the target explicitly receives a message.

In a general mode of operation, the parallel processes may be executed totally asynchronously to each other. Since the communication constructs are synchronization points for the scheduling of the processes, the programmer must be aware of the data dependencies among them. For complex massively parallel problems with a large multitude of parallel processes, it is practically impossible for humans to perform such a totally asynchronous scheduling. This makes the message-passing programming difficult.

To make message-passing programming feasible, restrictions must be imposed on the manner in which parallel processing takes place. A most helpful restriction is introduced by the SPMD model (SPMD: single program - multiple data). This model can be applied to data parallel problems in which one and the same basic code executes against partitioned data. Such programs execute in a loosely synchronous style with computation phases alternating with communication phases. During the computation phase, each process

---

[10] In general, we can talk about parallel program segments; they may be common heavyweight processes or teams of lightweight processes (threads) [130].

computes on its own portion of the data; during the communication phase, the processes exchange data using a message passing library.

As the programmer has full control over nearly all the parallel aspects, including data distribution, work distribution etc., there is greater opportunity to gain high performance. For example, the programmer can arrange the overlapping of computation, I/O and communication in some cases to optimize the code.

The basic concept of processes communicating through messages is well understood. Over the last ten years, substantial progress has been made in casting significant applications into this paradigm. Unfortunately, each vendor has implemented his own variant. More recently, several public-domain systems, for instance PARMACS [36], PVM [30] and Chameleon [133] have demonstrated that the message-passing programming model can be efficiently and portably implemented. This effort issued into specification of the Message Passing Interface Standard (MPI) standard [247, 132] that defines the user interface and functionality for a wide range of message-passing capabilities.

Existing I/O systems for parallel computers have tended to concentrate on providing access to file systems at a fairly low level; users often need to know details of disk-blocking and read-write caching strategies to get acceptable performance. Choosing the appropriate extensions to MPI (including new datatypes and datatype constructors) as well as handling the nonuniformity in access to the external devices is a challenging research area. The MPI-IO standardization effort [87] it is intended to leverage the widely accepted MPI for use in expressing parallel I/O activity. This use has the advantage of familiar syntax and semantics for programmers.

2.2.2 Data Parallel Programming Model

HPF and Vienna Fortran extended by concepts introduced in this book implement an *I/O oriented data parallel programming model* that is depicted in Fig. 2.4 (this kind of model presentation is partially motivated by [6]). This model supports a programming style that is hoped to be prevalent in programming I/O intensive scientific and engineering applications. Application programs are written using a sequential language to specify the computations on the data (using loops or array operations), and annotation to specify how large arrays should be distributed across internal memory of logical computational nodes and external memory of logical I/O nodes or directly across logical disks.

A parallelizing compiler then translates the source code into an equivalent SPMD program with message passing calls. Like in the MPI-IO approach [87], I/O is modeled as a specific kind of message passing. The computation is distributed to the parallel computational processes and I/O operations to the parallel I/O processes to match the specified data distributions. The compiler extracts some parameters about data distributions and data access

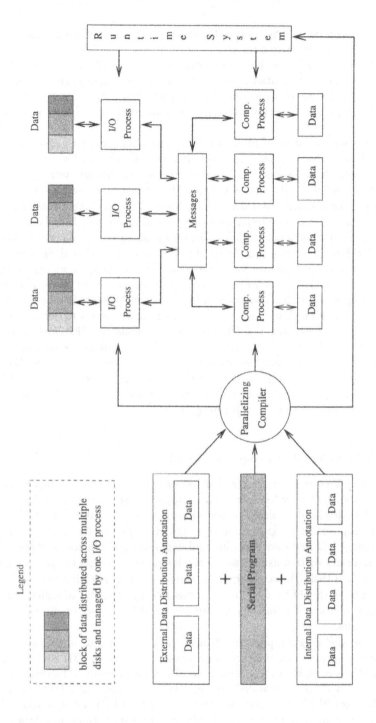

Fig. 2.4. I/O Oriented Data Parallel Programming Model.

patterns from the source program and passes them to the runtime system. Based on this information, the runtime system organizes the data and tries to ensure optimal load balance and high performance access to stored data. It should be noted that the distribution annotation is only assumed to provide the compiler and runtime systems with the user's preferences and hints for the final data layout rather than as commands.

2.2.3 Models Based on Coordination Mechanisms

Programming for DMSs can be made machine independent by *coordination languages*. A coordination language usually is a set of primitives for data distribution, process creation, synchronization and communication. The complete parallel programming language is obtained by taking a sequential host language (an ordinary computation language, e.g., Fortran) that provides data types and sequential control and augment it by the coordination language. The host language embodies some sequential computational model. A coordination language (e.g., Linda [69]) embodies some coordination model. It abstracts the relevant aspects of a program's parallel behaviour; e.g., it provides annotation for data distribution, operations to create computational activities and to support communication among them.

The coordination specification aims to provide sufficient information about the parallel structure of the program for the compiler to generate an efficient and correct code.

A computation and a coordination model can be integrated and embodied into a single programming language. SCL (Structured Coordination Language) [94] is an example of this approach.

BIBLIOGRAPHICAL NOTES

The relevant developments within the field of abstract parallel machine models are represented by papers [93, 184, 203, 239, 250, 238, 253, 253, 254, 262]. Shriver and Nodine [232] present an introduction to the data-transfer models on which most of the out-of-core parallel I/O algorithms are based, with particular emphasis on the Parallel Disk Model. Sample algorithms are discussed to demonstrate the paradigms used with these models. An introduction to I/O architectural issues in multiprocessors, with a focus on disk subsystems is provided by Kotz in [173]. Chen et al. [76] and Gibson [121] provide detailed surveys on disk arrays. Programming models for massively parallel computers are explicitly addressed in the conference proceedings [124, 125].

3. Vienna Fortran 90 and Its Extensions for Parallel I/O

This chapter describes *Vienna Fortran 90 (VF90)*, a data parallel language that gives the user full control over the crucial task of selecting a distribution of the program's data both in internal and external memory. VF90 is based on the IMPA machine model and provides support for the operations on files stored in a massively parallel storage system and the operations on out-of-core structures. The language extensions have been designed with a special focus on both alleviating the specification of both in-core and out-of-core parallel algorithms and the compiler's complex task of program optimization.

The part of VF90 supporting the development of computationally intensive in-core applications has been developed by Zima et al. (Fortran 77 extensions) [273] and Benkner [22]. Here, we only present the basic language features; a more detailed overview, and full description, including examples, can be found in [26, 22, 72, 273]. In this chapter, the presentation is mainly focused on the language extensions concerning parallel I/O. The formal definition of the VF90 I/O operations using the model of the denotational semantics can be found in [51].

3.1 Language Model

A VF90 program Q is executed on IMPA by running the code produced by the Vienna Fortran compiler, called the *Vienna Fortran Compilation System (VFCS)*. Computational tasks are executed on the computational nodes using the SPMD execution model. The I/O data management is handled by I/O nodes. The set of declared arrays of a program Q represents the *internal data space* A of Q. Thereby, scalar variables can be interpreted as one-dimensional arrays with one element.

Arrays from A are assigned to processor internal memories of the computational nodes. The way how the data is distributed does not influence the semantics of the program but may have a profound effect on the efficiency of the parallelized code. To achieve the efficient transfer of data between external and internal memory, the appropriate assignment of external data to disk memories of the I/O nodes is needed.

3.1.1 Index Domain and Index Mapping

First it is useful to present a few basic definitions used within the rest of this chapter.

Definition 3.1.1. Index Dimension

An index dimension is a non-empty, linearly ordered set of integers. It is called

- *a standard index dimension iff it can be represented in the form $[l : u]$, where $l \leq u$ and $[l : u]$ denotes the sequence of numbers $(l, l+1, \ldots, u)$,*
- *a normal index dimension[1] iff it is of the form $[1 : u]$,*
- *a regular index dimension iff it can be constructed from a normal index dimension through the application of a linear function of the form $i * s + c$, $abs(s) > 1$, to the elements of a normal index dimension and can be represented in the form $[l : u : s]$, where for $l \leq u$ and $s > 0$, $[l : u : s]$ denotes the sequence of numbers $(l, l + s, \ldots, u)$, and for $l > u$ and $s < 0$ $[l : u : s]$ denotes the sequence of numbers $(l, l - s, \ldots, u)$,*
- *an irregular index dimension in all other cases.*

Definition 3.1.2. Index Domain

1. *Let D_i represent an index dimension. Then an index domain of rank n, denoted by $\mathbf{I}$, can be represented in the form $\mathbf{I} = \mathbf{X}_{i=1}^n D_i$, where $\mathbf{X}$ denotes the Cartesian product. $\mathbf{I}$ is called*
   - *a standard index domain iff all D_i are standard index dimensions,*
   - *a normal index domain iff all D_i are normal index dimensions,*
   - *a regular index domain iff at least one index dimension D_i is a regular index dimensions, and all others are standard index dimensions.*
   - *a irregular index domain iff at least one index dimension D_i, $1 \leq i \leq n$, is an irregular index dimension.*
   In the following, let $\mathbf{I}$ denote an index domain of rank n, and i integer number with $1 \leq i \leq n$.
2. *The projection of $\mathbf{I}$ to its i-th component, D_i, is denoted by $\mathbf{I}_i$.*
3. *$| D_i |$ is the extent of dimension i (the number of elements along that dimension).*
4. *The shape of $\mathbf{I}$ is defined by $shape(\mathbf{I}) := (| D_1 |, \ldots, | D_n |)$.*

Definition 3.1.3. Index Mapping

An index mapping ι from an index domain $\mathbf{I}$ to an index domain $\mathbf{J}$ is a total function that maps each element of $\mathbf{I}$ to a non-empty set of elements of $\mathbf{J}$.

$$\iota : \mathbf{I} \rightarrow \mathcal{P}(\mathbf{J}) - \{\phi\},$$

where $\mathcal{P}$ denotes the power set.

---

[1] A normal index dimension is a also standard index dimension.

As will be seen later, index mappings are the basis for modeling data distributions.

Definition 3.1.4. Array Index Domain

Let in the following $\mathcal{A}$ denote the set of all data objects (scalars and arrays) declared in the program. Assume that $A \in \mathcal{A}$ is an arbitrary declared array.

1. *A is associated with a standard index domain, $\mathbf{I}^A$. All attributes of $\mathbf{I}^A$, such as rank and shape, are applied to A with the same meaning as specified in Definition 3.1.2.[2]*
2. *$\mathcal{E}^A$ is the set of elements of A. The elements are scalar objects.*
3. *$index^A : \mathcal{E}^A \rightarrow \mathbf{I}^A$ is a function establishing a one-to-one correspondence between $\mathcal{E}^A$ and $\mathbf{I}^A$. For every array element $e \in \mathcal{E}^A$, $index^A(e)$ is called the index of e.*

3.1.2 Index Domain of an Array Section

Let us consider an array A of rank n with index domain $\mathbf{I}^A = [l_1 : u_1, \ldots, l_n : u_n]$. The index domain of an array section $A' = A(ss_1, \ldots, ss_n)$ is of rank n', $1 \le n' \le n$, iff n' subscripts ss_i are section subscripts (subscript triplet or vector subscript) and the remaining subscripts are scalars. Due to the Fortran 90 conventions the index domain of an array section can always be classified as a normal index domain. For example, the index domain of the array section $A(11 : 100 : 2, 3)$ is given by $[1 : 45]$.)

The extended index domain of the array section A', denoted by $\bar{\mathbf{I}}^{A'}$, is given by $\bar{\mathbf{I}}^{A'} = [ss_1', \ldots, ss_n']$ where $ss_i' = ss_i$, if ss_i is a section subscript, and $ss_i' = [c_i : c_i]$ if ss_i is a scalar subscript that is denoted by c_i.

3.1.3 Data Distribution Model for Internal Memory

A major part of the Vienna Fortran extensions are aimed at specifying the alignment and distribution of the data elements. The underlying intuition for such mapping of data is as follows. If the computations on different elements of a data structure are independent, then distributing the data structure will allow the computation to be executed in parallel. Similarly, if elements of two data structures are used in the same computation, then they should be aligned so that they reside in the same processor memory. Obviously, the two factors may be in conflict across computations, giving rise to situations where data needed in a computation resides on some other processor. This data dependence is then satisfied by communicating the data from one processor to another. Thus, the main goal of mapping data onto processor memories is to increase parallelism while minimizing communication so that the workload across the processors is balanced.

---

[2] Whenever A is implied by the context, the superscript may be omitted. Analogous conventions hold for all similar cases.

Processor Arrays. In Vienna Fortran processors are explicitly introduced by the declaration of a processor array, by means of the PROCESSORS statement. These processors are abstractions of the IMPA nodes which fulfill computational tasks. The mapping of processor arrays to the physical machine is hidden in the compiler. Notational conventions introduced for arrays above can also be used for processor arrays, i.e., $\mathbf{I}^R$ denotes the index domain of a processor array R. More than one processor declaration is permitted in order to provide different *views* of the array; all others are *secondary processor arrays* or reshapes of the primary processor array. Correspondence between different processor arrays is established by means of the Fortran element order. The number of processors on which the program executes may be accessed by the intrinsic function $NP.

Distributions. A distribution of an array maps each array element to one or more processors which become the owners of the element, and, in this capacity, store the element in their local memory. We model distributions by functions, called distribution functions, between the associated index domains.

Definition 3.1.5. Distributions

Let $A \in \mathcal{A}$, and assume that R is a processor array. A distribution function of the array A with respect to R is defined by the the mapping:

$$\mu_R^A : \mathbf{I}^A \rightarrow \mathcal{P}(\mathbf{I}^R) - \{\phi\}$$

where $\mathcal{P}(\mathbf{I}^R)$ denotes the power set of $\mathbf{I}^R$.

Static Versus Dynamic Distributions. In Vienna Fortran, there are two classes of distributed arrays, *statically* and *dynamically* distributed arrays. The distribution of a statically distributed array may not change within the scope of its declaration. An array whose distribution may be modified within the scope of its declaration is called a dynamically distributed array and must be declared with the DYNAMIC attribute.

Alignments. By means of alignment it is possible to indirectly specify the distribution of an array by specifying its relative position with respect to another distributed array.

Definition 3.1.6. Alignments.

An alignment relationship of an array B with respect to an array A is established by means of an alignment function which is an index mapping from the index domain of array B to the index domain of array A. Such an alignment function is denoted by

$$\alpha_A^B : \mathbf{I}^B \rightarrow \mathcal{P}(\mathbf{I}^A) - \{\phi\}$$

A is called alignment source *array and B is called* alignee.

The distribution of an alignee can be constructed by means of the distribution of the corresponding alignment source array as follows:

Definition 3.1.7. Construction of a Distribution

If $\mathbf{I}^A, \mathbf{I}^B$, μ_R^B, and α_A^B are given as above, then μ_R^B can be constructed as follows: For each $\mathbf{i} \in \mathbf{I}^B$:

$$\mu_R^B(\mathbf{i}) := \bigcup_{\mathbf{j} \in \alpha(\mathbf{i})} \mu_R^A(\mathbf{j})$$

As a consequence, if index $\mathbf{i} \in \mathbf{I}^B$ is mapped via the alignment function α_A^B to an index $\mathbf{j} \in \mathbf{I}^A$, then it is guaranteed that the elements $B(\mathbf{i})$ and $A(\mathbf{j})$ are allocated to the same processor(s), regardless of the distribution of A.

Distribution Types. A *distribution type* specifies a class of distributions rather than individual distributions. It is determined by one distribution function or - in the case of composite distributions (see below) - a tuple of distribution functions together with the list of evaluated arguments. The application of a distribution type to an environment that specifies a processor array and data array (section) yields a distribution.

3.2 Language Constructs

3.2.1 Distribution Annotations

Distribution can be specified either directly, by means of the distribution function, or implicitly by referring to the distribution of another array. The direct distribution of a statically distributed array in VF90 must be specified within its declaration by providing the **DISTRIBUTED** attribute followed by a *distribution specification*. An optional *processor reference* after the keyword **TO**, specifies the processor array to which the distribution refers. If no processor reference is present, then data arrays are mapped by default to the primary processor array. Dynamically distributed arrays are usually associated with a distribution by execution of a **DISTRIBUTE** statement.

Due to Definition 3.1.5, distribution function for an array maps each array element to one or more processors, which store those elements in their local memory. A distribution function is called a *general* or *n-m distribution function* if it defines the distribution of an n-dimensional array with respect to an m-dimensional processor array. An important class of distributions are *composite distributions*. A composite distribution is specified by describing the distribution of each dimension separately, without any interaction of dimensions, by means of *dimensional distribution functions*. A dimensional distribution maps each index of an array dimension to one or more indices of a processor array dimension. To specify regular dimensional distributions[3], intrinsic functions BLOCK, BLOCK(M), GENERAL_BLOCK(*block-lengths*) and CYCLIC(L) and the elision symbol ":" indicating that the corresponding dimension is not distributed, may be used[4].

We illustrate regular distributions and replications and the annotation syntax by means of an example (Fig. 3.1(a)). The PROCESSORS declaration introduces $R1$ as a one-dimensional processor array with four elements. The annotations attached to the declarations specify a regular *block distribution* for A, *replication* for array B, and a *general block distribution* for C. The last one distributes an array dimension to a processor dimension in arbitrarily sized portions. Array C is partitioned into 4 blocks of lengths 1,5,5, and 1. The annotation in the declaration of D specifies a block-wise distribution of its columns. Since in this example $R1$ is the only processor array specified, it is always implied and can be omitted in the annotation. The resulting data-layout is illustrated in Fig. 3.1(c).

The corresponding HPF notation for arrays A, B and D is introduced in Fig. 3.1(b). There is no support for a GENERAL_BLOCK distribution in HPF.

The specifications introduced in Fig. 3.1(a) can be rewritten into the corresponding Vienna Fortran 77 notation in the following way:

```
PROCESSORS R1(1:4)
REAL A(12) DIST ( BLOCK ) TO R1
REAL B(5) DIST (:)
REAL C(12) DIST ( GENERAL_BLOCK (1,5,5,1))
REAL D(100,100) DIST (:, BLOCK )
```

Remark: In this book, some examples which are not included in figures are marked by ▔▔▔▼ and ▁▁▁▲.

By default, an array which is not explicitly distributed is replicated to all processors.

---

[3] A dimensional distribution is referred to as *regular* if an array dimension is partitioned into disjoint parts in such a way that each part can be represented by means of a regular section.

[4] These distribution types are also provided by HPF, however, symbol "*" is used instead of ":" in HPF.

(a) Vienna Fortran 90 annotation

```
PROCESSORS :: R1(1:4)
REAL, DISTRIBUTED (BLOCK) TO R1 :: A(12)
REAL, DISTRIBUTED (:) :: B(5)
REAL, DISTRIBUTED (GENERAL_BLOCK (1,5,5,1)) :: C(12)
REAL, DISTRIBUTED (:, BLOCK) :: D(100,100)
```

(b) HPF annotation

```
!HPF$ PROCESSORS R1(1:4)
      REAL  A(12), B(5), D(100,100)
!HPF$ DISTRIBUTE (BLOCK) ONTO R1 :: A
!HPF$ DISTRIBUTE (*) ONTO R1 :: B
!HPF$ DISTRIBUTE (*, BLOCK) ONTO R1 :: D
```

(c) Data Layout

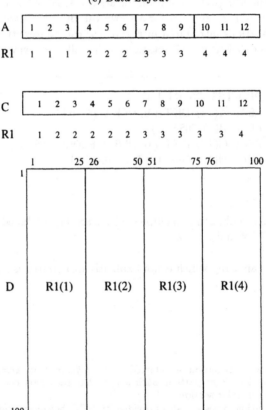

Fig. 3.1. Basic Distributions

3.2.2 Alignment Annotations

Implicit distributions begin with the keyword ALIGNED and require both the target array and a source array (so called because it is the source of the distribution). An element of the target array is distributed to the same processor as the specified element of the source array, which is determined by evaluating the expressions in the source array description for each valid subscript of the target array. Here, II and JJ are bound variables in the annotation, and range in value from 1 through 80.

INTEGER, ALIGNED IM(II,JJ) WITH D(JJ,II+10) :: IM(80,80)

3.2.3 Dynamic Distributions and the DISTRIBUTE Statement

By default, the distribution of an array is static. Thus it does not change within the scope of the declaration to which the distribution has been appended. The keyword DYNAMIC is provided to declare an array distribution to be dynamic. This permits the array to be the target of a DISTRIBUTE statement. A dynamically distributed array may optionally be provided with an initial distribution in the manner described above for static distributions. A range of permissible distributions may be specified when the array is declared by giving the keyword RANGE and a set of explicit distributions. If this does not appear, the array may take on any permitted distribution with the appropriate dimensionality during execution of the program. Finally, the distribution of such an array may be dynamically connected to the distribution of another dynamically distributed array in a specified fixed manner. This is expressed by means of the CONNECT keyword. Thus, if the latter array is redistributed, then the connected array will automatically also be redistributed. For example,

```
REAL, DYNAMIC, &
    RANGE (( BLOCK , BLOCK ),( CYCLIC (5), BLOCK )) :: F(200,200)
REAL, DYNAMIC, CONNECT (=F) :: G(200,200), H(200,200)
```

An array with the DYNAMIC attribute may be given a new distribution by executing a DISTRIBUTE statement. If the NOTRANSFER attribute is given for an array within a distribute statement, no physical transfer of data takes place: only the access function of the array is changed according to the new distribution. In the following example, both arrays A and B are redistributed with the new distribution CYCLIC(10).

DISTRIBUTE A, B :: (CYCLIC(10))

3.2.4 Indirect Distribution

VF90 provides indirect distributions to support irregular applications. Indirect distribution, defined by means of the intrinsic function INDIRECT and a mapping array, allows arbitrary mappings of array elements to processors. Mapping arrays must be integer arrays with semantic constraints on their values and may be associated with a distribution in the usual way. The distribution function INDIRECT may be used as a dimensional distribution function to specify the distribution of a single array dimension or as a general distribution function.

```
PARAMETER :: N = 10, M = 8
PROCESSORS R2(4,2), R(8)
INTEGER, PARAMETER :: MAP1(N) = (/ 1,2,3,4,4,4,5,6,7,8 /)
INTEGER, PARAMETER :: MAP2(N) = (/ 1,2,3,4,4,4,3,2,1,1 /)
INTEGER, PARAMETER :: MAP3(N,M) =              &
    RESHAPE( FILEVALS ('dist_C.map', N*M, 'INTEGER'), (/N,M/))
REAL, DISTRIBUTED( INDIRECT (MAP1), :) TO R :: A(N,N)
REAL, DISTRIBUTED( INDIRECT (MAP2), BLOCK) TO R2 :: B(N,N)
REAL, DISTRIBUTED( INDIRECT (MAP3)) :: C(N,M)
```

Fig. 3.2. Examples of Irregularly Distributed Arrays

If INDIRECT represents a dimensional distribution function, then the mapping array must be 1-dimensional and must have the same size as the corresponding array dimension. Correspondence between the indices of the array dimension and the elements of the mapping array is established by means of the Fortran array element order. Each element of the mapping array must represent a valid index of the associated index dimension of the corresponding processor array, that means the processor array structure must be taken into account (look at arrays A and B in Fig. 3.2 and 3.3). If INDIRECT is used as a general distribution function, then the mapping array must be shaped conform with the array distributed. Correspondence between the indices of the array dimension and the elements of the mapping array is established by means of the Fortran array element order. Each element of the mapping array must represent a valid index of the corresponding linearized processor array (look at array C in Fig. 3.2 and 3.3).

Examples of indirect distributions are shown in Fig. 3.2. For keeping figures as clear as possible, we use indices of the processor array R only. The relationship between the elements of $R2$ and those of R is apparent. The layouts of elements of arrays are depicted in Fig. 3.3.

Arrays A and B are indirectly distributed in the first dimension according to the contents of arrays $MAP1$ and $MAP2$, respectively. The second dimen-

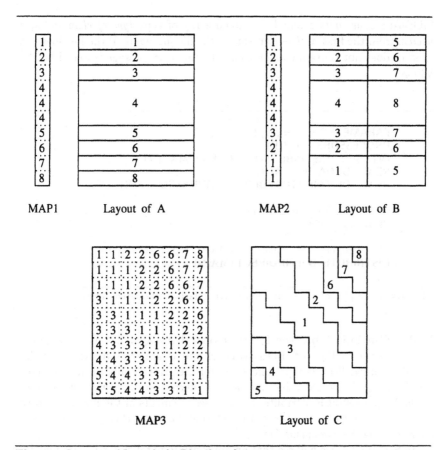

Fig. 3.3. Layouts of Irregularly Distributed Arrays

sion of A is not distributed, while the second dimension of B has got the BLOCK distribution. The mapping arrays *MAP1* and *MAP2* are initialized by N integer constants specified by the *array_constructors*.

Array C is distributed applying the general INDIRECT distribution function according to the contents of the 2-dimensional array *MAP3*. A new intrinsic function FILEVALS is utilized for initializing the mapping array MAP3. FILEVALS constructs a 1-dimensional array from the elements of an existing file *"dist_C.map"* that is handled by the VFCS as a special source file. In this context we will call such files as *static mapping files*. Generally, arguments to FILEVALS are :

file_name - name of the file (character literal constant)
n_elem - number of elements to be processed from the *file_name* (integer specification expression)
type_of_elem - type of the file elements (character literal constant)

FILEVALS constructs a one-dimensional array of type *type_of_elem* and size *n_elem* from the first *n_elem* elements of the file *file_name*. If the file does not exist or does not contain enough elements, an error message is issued by the VFCS.

```
PARAMETER :: N = ... , NP = ...
PROCESSORS :: R(NP)
INTEGER, DISTRIBUTED ( BLOCK ) :: MAP(N)
REAL , DYNAMIC :: A(N)
REAL , DISTRIBUTED ( BLOCK ), DYNAMIC :: B(N)
     . . .
OPEN (iu, file='dist_AB.map')
READ (iu,*) MAP
     . . .
DISTRIBUTE A :: ( INDIRECT (MAP), NOTRANSFER (A))
DISTRIBUTE B :: ( INDIRECT (MAP))
```

Fig. 3.4. Examples of Redistribution of Irregularly Distributed Arrays

Redistribution of Indirectly Distributed Arrays. Let us suppose the example given in Fig. 3.4. Both arrays A and B are redistributed with the distribution INDIRECT(MAP), however, for the array A only the access function is changed, the old values are not transferred to the new locations, due to the NOTRANSFER attribute. In this case, elements of the mapping array are assigned by values read from the external file *dist_AB.map* which we will refer to as dynamic mapping file.

In Section 4.4, we will see that an appropriate parallel partitioner may be used for the computation of the mapping array at runtime ([214]).

3.2.5 Procedures

Declarations of dummy arguments and local arrays in procedures can be annotated with a distribution in precisely the same way as declarations of arrays in the main program can be. In addition, a dummy argument may inherit its distribution from the actual argument. This enables the user to write procedures which will accept actual arguments with a variety of distributions. Within the procedure, the distribution of the actual argument may be used, or a specific distribution may be enforced, as required. The specification and compiling of procedure interfaces with explicit (non-inherited) distributions is discussed in [25, 22]. Here, we focus on implicit indirect distributions.

Calls to procedures must leave the distribution of a statically distributed actual array unchanged. Thus on procedure exit, an implicit redistribution has to be performed to restore the original distribution. Dynamically distributed actual arrays may, however, have a modified distribution after a procedure call if the corresponding dummy arrays have been redistributed inside the procedure. Restoring the original distribution can always be enforced by specifying the RESTORE attribute, either in the formal argument specification or with the actual argument at the point of the call. This scheme for argument passing is illustrated in example given in Fig. 3.5.

```
DISTRIBUTE A, B :: ( INDIRECT (MAP1))
CALL SUB(A :: RESTORE , B :: RESTORE , MAP2)
       ...
SUBROUTINE SUB(X, Y, M)
     REAL , DISTRIBUTED (∗), DIMENSION (N) :: X, Y
     INTEGER , DISTRIBUTED (∗), DIMENSION (N) :: M
     REAL , DISTRIBUTED ( INDIRECT (M) :: W(N)
       ...
     DISTRIBUTE  X :: ( INDIRECT (M))
END SUBROUTINE SUB
```

Fig. 3.5. Examples of Irregularly Distributed Arguments

In this code, dummy argument arrays X and Y of subroutine SUB inherit dynamical irregular distributions of actual argument arrays A and B, respectively. The third dummy argument array M inherits the $MAP2$'s distribution. It is guaranteed that the distributions of A and B are not changed by the call of SUB.

3.2.6 FORALL Loop

Irregular computations can be described by explicitly data parallel loops - forall loops. A precondition for the correctness of a forall loop is the absence of loop-carried dependences except accumulation type dependences. The general form of the forall loop is introduced in the following:

FORALL $(I_1=\text{sec}_1, \dots , I_n=\text{sec}_n)[\textit{Work\_distr\_spec}]$
    forall-body
END FORALL

where for $1 \leq j \leq n$, I_j are index names, and sec_j are subscript triplets of the form (l_j, u_j, s_j) which must not contain references to any index name. If the value of s_j equals one, s_j can be omitted. If j equals one, the header parenthesis can also be left out. The range of the index variable I_j is specified by the corresponding section sec_j. The loop iterates over the Cartesian product of the individual ranges.

Work_distr_spec denotes an optional work distribution specification. If no specification is introduced by the user, it is derived automatically by the compiler. Work distribution assigns loop iterations to processors for execution.

```
FORALL i = 1, N ON OWNER (E1(i))
        B(E1(i)) = B(E1(i)) + f(A(E1(i)), A(E2(i)))
END FORALL

FORALL i = 1, N ON  P1D(MOD(i,NP1)+1)
        X(i) = Y(i) + Y(LC(i,1)) - Y(LC(i,2))
END FORALL
```

Fig. 3.6. Work Distribution Based on Data Ownership (Upper), and on Direct Naming the Target Processor (Lower)

In the first forall loop in Fig. 3.6(upper), the work distribution is specified using the intrinsic function **OWNER**, which returns the home processor of its argument. Thus, the above *on clause* specifies that the ith iteration of the forall loop should be executed on the processor which owns the array element $E1(i)$. The processor may also be named directly as shown in Fig. 3.6(lower) In this case, the value of the index expression $MOD(i,NP1)+1$ determines the processor which will execute the iteration i.

Furthermore, if the iteration space of the loop is rectilinear, the distribution of the loop iterations may be specified by means of a distribution specification in much the same way as for the distribution of data arrays. For example:

FORALL $((i = L_1,U_1), (j = L_2,U_2))$ DISTRIBUTED (CYCLIC, BLOCK) TO P2D

3.2.7 Reduction Operators

As indicated before, forall loops in Vienna Fortran are not allowed to have inter-iteration dependences. One consequence of this restriction is that no scalar variable can be modified in a forall loop. Vienna Fortran allows reduction operators to represent operations such as summation across the iterations of a parallel loop. The general form of the reduction statement is as follows:

REDUCE (*reduction-op*, *target-var*, *expression*)

The effect of the statement is to accumulate the values of the expression onto the target variable.

The example in Fig. 3.7(a) illustrates the summing values of a distributed array. In this loop, the reduction statement along with the reduction operator

ADD is used to sum the values of the distributed array A onto the variable S. Along with ADD, Vienna Fortran provides MULT, MAX, and MIN as reduction operators. The accumulating values onto a distributed array is shown in Fig. 3.7(b). In each iteration, elements of A and B are multiplied, and then the result is used to increment the corresponding element of X. In general, all the arrays X, A, B, $I1$, $I2$, $I3$ can be distributed.

(a) Summing values of a distributed array

```
S = 0.0
FORALL i = 1,N  ON OWNER (A(X(i)))
        REDUCE ( ADD , S, A(X(i)))
END FORALL
```

(b) Accumulating values onto a distributed array

```
FORALL i = 1,N  DISTRIBUTED ( BLOCK )
        REDUCE ( ADD ,X(I1(i)),A(I2(i))*B(I3(i)))
END FORALL
```

Fig. 3.7. Reduction Types

3.2.8 User-Defined Distribution Functions

An intended distribution (or class of them) can be bound to a name by means of a *user-defined distribution function (UDDF)* which may have arguments in order to allow for parameterization.

UDDFs provide a facility for extending the set of intrinsic mappings defined in the language in a structured way. The specification of a UDDF establishes a mapping from (data) arrays to processor arrays, using a syntax similar to Fortran functions. UDDFs may contain local data structures and executable statements along with at least one *distribution mapping statement* which maps the elements of the target array to the processors.

The use of UDDFs will be illustrated in examples in Section 3.3.

3.2.9 Support for Sparse Matrix Computation

Requirements for sparse matrix computations arise in many problem areas, such as finite elements, molecular dynamics, matrix decomposition, and image reconstruction.

Vienna Fortran introduces additional language functionality to allow the efficient processing of sparse matrix codes [249]. The approach is based on new methods for the representation and distribution of sparse matrices which

$$\begin{pmatrix}
0 & 53 & 0 & 0 & 0 & 0 & 0 & 0 \\
0 & 0 & 0 & 0 & 0 & 0 & 21 & 0 \\
19 & 0 & 0 & 0 & 0 & 0 & 0 & 16 \\
0 & 0 & 0 & 0 & 0 & 72 & 0 & 0 \\
0 & 0 & 0 & 17 & 0 & 0 & 0 & 0 \\
0 & 0 & 0 & 0 & 93 & 0 & 0 & 0 \\
0 & 0 & 0 & 0 & 0 & 0 & 13 & 0 \\
0 & 0 & 0 & 0 & 44 & 0 & 0 & 19 \\
0 & 23 & 69 & 0 & 37 & 0 & 0 & 0 \\
27 & 0 & 0 & 11 & 0 & 0 & 64 & 0
\end{pmatrix}$$

Fig. 3.8. Sample Sparse Matrix A(10,8) with 16 Non-Zero Elements

form a powerful mechanism for storing and manipulating sparse matrices able to be efficiently implemented on massively parallel machines. The special nature of these new language features leads to compile time and runtime optimizations which save memory, reduce the amount of communication and generally improve the runtime behavior of the application.

Representing Sparse Matrices on DMSs. Vienna Fortran constructs a representation of a sparse matrix on a DMS by combining a standard representation used on sequential machines with a data distribution in the following way.

A *distributed representation (d-representation)* is determined by two components, a *sequential representation (s-representation)* and a data distribution. The s-representation specifies a set of data structures, which store the data of the sparse matrix and establish associated access mechanisms on a sequential machine. The distribution determines, in the usual sense, a mapping of the array to the processors of the machine.

For the following, assume a processor array $Q(0 : X - 1, 0 : Y - 1)$ and a data array $A(1 : n, 1 : m)$.

We describe a frequently used s-representation called the *Compressed Row Storage (CRS)* representation. This representation is determined by a triple of vectors, (DA, CO, RO), which are respectively called the *data*, *column*, and *row* vector. The data vector stores the non-zero values of the matrix, as they are traversed in a row-wise fashion. The column vector stores the column indices of the elements in the data vector. Finally, the row vector stores the indices in the data vector that correspond to the first non-zero element of each row (if such an element exists).

For example, consider the sparse matrix A shown in Fig. 3.8. The data, column and row vectors for A are shown in Fig. 3.9.

Another s-representation, *Compressed Column Storage (CCS)*, is similar to CRS, but based on an enumeration which traverses the columns of A rather than the rows.

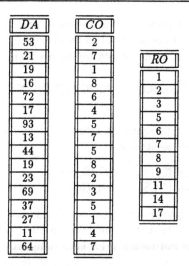

Fig. 3.9. CRS Representation for the Matrix A

A *d-representation* for A results from combining a data distribution with the given s-representation. This is to be understood as follows: The data distribution determines for each processor an associated set of local elements called the *local segment* which (under appropriate constraints) is again a sparse matrix. Now the local segment of each processor will be represented in a manner analogous to the s-representation employed in the sequential code. More specifically, DA, CO, and RO are automatically converted to sets of vectors DA^p, CO^p, and RO^p, for each processor p. Hence the parallel code will save computation and storage using the very same mechanisms that were applied in the original program. In practice we need some additional global information to support exchanges of sparse data with other processors.

Below we describe two strategies to represent and distribute sparse matrices on multiprocessors.

Multiple Recursive Decomposition (MRD).

MRD is a data distribution that generalizes the Binary Recursive Decomposition (BRD) of Berger and Bokhari [31] to an *arbitrary* two-dimensional array of processors. This distribution defines the local segment of each processor as a rectangular matrix which preserves neighborhood properties and achieves a good load balance.

BRS and BCS.

The BRS and BCS representations are based upon CYCLIC distributions: both dimensions of A are assumed to be distributed cyclically with block length 1, as specified by the annotation in the Vienna Fortran declaration

REAL, DISTRIBUTED (CYCLIC , CYCLIC) TO Q(0:X-1,0:Y-1) :: A(N,M)

We obtain the desired representation schemes by combining this distribution with two kinds of s-representation:

– Block Row Scatter (BRS) – using CRS, and
– Block Column Scatter (BCS) – using CCS.

The BRS and BCS representations are good choices for irregular algorithms with a gradual decrease or increase in the workload, and for those where the workload is not identical for each non-zero element of the sparse matrix. Many common algorithms are of this nature, including sparse matrices decompositions (LU, Cholesky, QR, WZ), image reconstruction (Expectation Maximization algorithm), least Square Minimum, iterative methods to solve linear systems (Conjugate and Biconjugate Gradient, Minimal Residual, Chebyshev Iteration, and others [20]) and eigenvalue solvers (Lanczos algorithm).

For example, the BRS representation of sparse matrix A in Fig. 3.8 is given in Fig. 3.10.

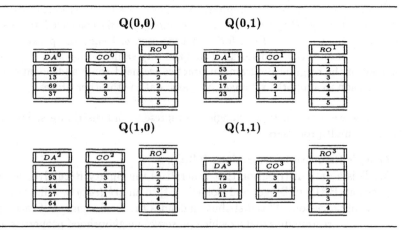

Fig. 3.10. BRS Representation of A to Q(0:1,0:1)

Vienna Fortran Extensions for the Support of Sparse Matrix Computations. We discuss here a small number of new language features for

Vienna Fortran (and, similarly, HPF) that support the sparse representations discussed above.

When the data are available at the beginning of execution, the original matrix is read from file and the d-representation can be constructed at compile-time. Otherwise, it has to be computed at runtime. Temporary and persistent sparse matrices can be stored directly in the d-representation; this issue will be discussed in Section 3.3.

We must take account of the fact that the data is accessed according to sparse structures in the source code. We require these auxiliary structures to be declared and used according to one of the representations known to the compiler. The compiler uses this information to construct the local data sets as described in the previous section.

This requires us to give the following information to the compiler:

- The *name, index domain*, and *element type* of the sparse matrix must be declared. This is done using regular Fortran declaration syntax; it creates an array resembling an ordinary Fortran array, but without the standard memory allocation.
- An annotation must be specified which declares the array as being SPARSE and provides information on the representation of the array. This includes the names of the auxiliary vectors in the specific order Data, Column and Row, which shall not be declared in the program.
- The keyword DYNAMIC is used in a manner analogous to its meaning in Vienna Fortran: if it is specified, then the d-representation will be determined dynamically, as a result of executing a DISTRIBUTE statement. Otherwise, the sparsity structure of the matrix is statically known, and thus all components of the d-representation (possibly excepting the actual non-zero values of the matrix) can be constructed in the compiler. Often, this information will be contained in a file whose name will be indicated in this annotation.

Sparse Matrix Product. One of the most important operations in matrix algebra is the matrix product. We present in Fig. 3.11 this algorithm expressed in Vienna Fortran and extended by the new sparse annotations. Both CCS and CRS representations are used here: while CRS is more suitable for the traversal of A, CCS is more appropriate for B in computing the product $C = A.B$.

3.3 Controlling Parallel I/O Operations

The above concepts related to the data distribution onto computational processors may be extended to the mapping of data onto external devices.

Array elements are stored on disks by means of files. The I/O profile of many scientific applications running on parallel systems is characterized

```
      PARAMETER (X=4,Y=4)
      PROCESSORS Q(0:X-1,0:Y-1)
      PARAMETER (NA=1000, NB=1000, NC=1000)
      REAL, DISTRIBUTED(CYCLIC,CYCLIC) :: C(NA,NC)
      INTEGER I, J, K

C  A uses Compressed Row Storage Format.
      REAL, DYNAMIC, SPARSE(CRS(AD,AC,AR)) :: A(NA, NB)
C  B uses Compressed Column Storage Format.
      REAL, DYNAMIC, SPARSE(CCS(BD,BC,BR)) :: B(NB, NC)
      ...
C — Read A and B
      ...
      DISTRIBUTE A,B :: (CYCLIC, CYCLIC)

C — Initialization of (dense) matrix C
      FORALL I = 1,NA
        FORALL J=1,NC
        C(I,J) = 0.0
        END FORALL
      END FORALL

C — Computation of the product.
      FORALL I = 1,NA
        FORALL K = 1,NC
        DO J1 = 1, AR(I+1)-AR(I)
          DO J2 = 1, BC(K+1)-BC(K)
          IF (AC(J1) .EQ. (BR(J2)) THEN
            C(I,K) = C(I,K)+AD(AR(I)+J1)*BD(BC(K)+J2)
          END DO
        END DO
        END FORALL
      END FORALL
      END
```

Fig. 3.11. Vienna-Fortran Sparse Matrix Product

by very large files which store array data which are mostly read in certain application classes and written in other ones. In many cases, the file data is distributed among the application processes, such that each reads a certain part or several parts, and all together read the whole file.

3.3.1 The File Processing Model

In typical computational science applications six types of I/O can be identified ([117]): (1) input, (2) debugging, (3) scratch files, (4) checkpoint/restart, (5) output, and (6) accessing out-of-core structures. Types (3), (4) and in some phases (6) too, do not contribute to the communication with the environment of the processing system. Therefore, the data they include may be stored on external devices of the parallel I/O system as *parallel files*, whose layout may be optimized to achieve the highest I/O data transfer rate. In the Vienna project [59], this I/O functionality is implemented by an advanced runtime system called VIPIOS (VIenna Parallel I/O System). Other I/O types include files that must have the standard format – *standard files*. The flow of I/O data in a typical application processing cycle is depicted in Fig. 3.12. Standard files can be read into or produced from either computational nodes of a distributed memory system (DMS) or VIPIOS. The transfer latency can partially or totally be hidden by computation. Parallel I/O operations do not appear to the user very different from sequential ones. The general logical user view of stored values in a VIPIOS file corresponds to the conventional sequential Fortran file model. The user need not to explicitly take care of the physical layout of the files; he or she may only provide the compiler and VIPIOS with information how the stored data will be used in the future. Like the NSL UniTree system [158], VIPIOS provides users with transparent access to virtually unlimited storage space. Files may automatically migrate to less expensive storage according to a programmable migration strategy.

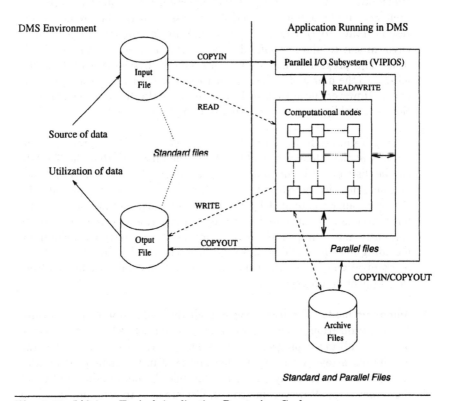

Fig. 3.12. I/O in a Typical Application Processing Cycle

3.3.2 User's Perspective

Various categories of users want to use the high-performance I/O. Advanced users with experience in high-performance computing are ready to work with low level programming features to get high performance for their applications. On the other hand, many of the users are specialists only in a specific application area and have no interest in penetrating into the secrets of programming massively parallel I/O subsystems. However, they may utilize their application domain specific knowledge and provide the programming environment with information about file access profiles that can be utilized in the optimization process. The third category is interested in high performance only and wants to work with I/O in the same way as in a sequential programming environment.

With regard to the users' aspects mentioned above, the main objective of our language proposal is to support a wide spectrum of users by enabling them to work with parallel I/O at several abstraction levels, which are outlined below.

1. Prescriptive mode. The user gets the opportunity to explicitly express his preference for layout of data on disks and assigning the data management responsibility to I/O nodes. This results in advising the programming environment to place specific data at specific I/O nodes and on specific disks. This programming style presents a more physical view of the parallel I/O to the user and makes it easier to write efficient hardware-specific codes. However, note that such an external data distribution specification is not obligatory for the programming environment.

2. Descriptive mode. In the form of *hints*, the user tells the programming environment the intended use of the file in the future, so that the file data can be distributed in an "optimum" way across the disks and I/O nodes and be dynamically redistributed if needed. The knowledge provided by the user can be extended by the results of program analysis. This information can be employed to organize the physical storage of the file efficiently and to implement high performance data retrieval techniques, including data prefetching and asynchronous parallel data retrieval and reorganization of files. All of the implementation details are hidden from the application programmer.

3. Automatic mode. The user delegates all the responsibility for the program optimization to the programming environment whose decisions are based on advanced program analysis.This approach can be used if the distributed arrays are being written and read in the same program, for example as scratch arrays, then the compiler knows the distribution of the target array. However, in general, files are used to communicate data between programs. In such situations, the compiler and runtime system

do not have any information about the distribution of the target array and hence will have to cooperate to determine the best possible layout for writing out the data, which will be system-dependent and based on heuristics.

In some cases, the user may prefer to use a combination of the above three modes.

3.3.3 Data Distribution Model for External Memory

This subsection proposes to expand the mapping model introduced in Section 3 to include files.

The File Mapping Concept. Through distribution of the file records onto disks, we indirectly achieve the distribution of data that is written to the file.

FILEGRID F(2,2,*)
OPEN (iu1, FILE = '/usr/example', FGRID = F)

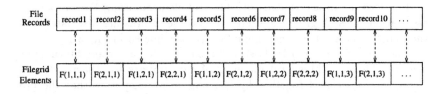

Fig. 3.13. Association Between File Records and Filegrid Elements.

A file can be viewed as a one-dimensional array of records whose lower bound equals 1 and the upper bound may be infinite when the number of records to be stored in it is in general unknown. These records can be mapped to I/O nodes and disks in a manner analogous to the mapping of data array to computational processors.

Files are not first class objects in Fortran and therefore, there is no direct way to explicitly specify their layout. Therefore, we introduce a new class of objects with the attribute FILEGRID being, essentially, named files. The attribute FILEGRID is similar to FILEMAP introduced by Marc Snir [236]. An actual file is associated with a filegrid in an OPEN statement by means of a new optional specifier FGRID. Filegrids provide one- or multi-dimensional views of files, giving full flexibility in the types of mapping that can be specified. In the case that a filegrid provides an abstraction of a file with an unknown size, the filegrid can be thought of as an assumed-size array of records.

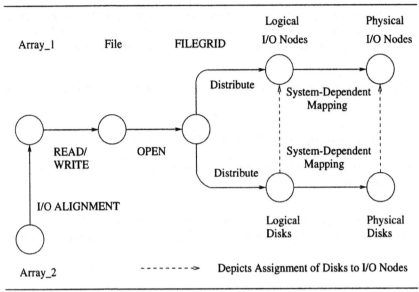

Fig. 3.14. Array and File Mapping.

File records are associated with file grid elements using Fortran storage association rules. A graphical sketch of how file records are associated with the elements of a three-dimensional file grid is depicted in Fig. 3.13.

The diagram in Fig. 3.14 illustrates the integrated array and file mapping concept. There is three-level mapping of data objects to abstract devices. Arrays are associated together by I/O alignment suggesting that the programming environment makes placement of elements of *Array_2* dependent on placement of elements of *Array_1*; *Array_1* is associated with a file through a write or read operation; a file is associated with a filegrid in an **OPEN** statement; the filegrid is then distributed onto rectilinear arrangements of logical disks and logical I/O nodes. The mapping of abstract I/O nodes and disks to their physical counterparts and the assignment of disks to I/O nodes is system-dependent.

The programmer working in the prescriptive mode is directly interfaced with this model. However, the programming environment always applies the model when handling data placement issues.

I/O Node Arrays and Disk Arrays. I/O node arrays are declared by means of the **IONODES** statement, and disk arrays by means of the **DISKS** statement. The notational conventions introduced for data arrays and processor arrays in Section 3.2 can also be used for I/O node and disk arrays, i.e., $\mathbf{I}^M$ and $\mathbf{I}^D$ will denote the index domains of an I/O node array M and a disk array D, respectively. Each array of disks may be decomposed into subarrays, each of them associated with an appropriate I/O node.

Definition 3.3.1. Association between disks and I/O nodes
*Let D be a disk array and M be an I/O node array. An association function
of the array D with respect to M is defined by the the mapping:*

$$\eta_M^D : \mathbf{I}^D \rightarrow \mathcal{P}(\mathbf{I}^M) - \{\phi\}.$$

where $\mathcal{P}(\mathbf{I}^M)$ denotes the power set of $\mathbf{I}^M$.

Data Distribution onto I/O Node Arrays and Disk Arrays. Using
the filegrid concept, it is possible to model the distribution of data arrays
onto disk arrays and I/O node arrays.

Definition 3.3.2. Distributions onto I/O Node Arrays
*Let $A \in \mathcal{A}$, and assume that M is an I/O node array. A distribution function
of the array A with respect to M is defined by the the mapping:*

$$\beta_M^A : \mathbf{I}^A \rightarrow \mathcal{P}(\mathbf{I}^M)$$

where $\mathcal{P}(\mathbf{I}^M)$ denotes the power set of $\mathbf{I}^M$.

Definition 3.3.3. Distributions onto Disk Arrays
*Let $A \in \mathcal{A}$, and assume that D is a disk array. A distribution function of the
array A with respect to D is defined by the the mapping:*

$$\gamma_D^A : \mathbf{I}^A \rightarrow \mathcal{P}(\mathbf{I}^D).$$

where $\mathcal{P}(\mathbf{I}^D)$ denotes the power set of $\mathbf{I}^D$.

A portion of A mapped to an I/O node is distributed onto the appropriate
disk subarray associated with this node in the following way: *Let $\bar{\mathbf{I}}^{A_i}$ denote
an extended index domain of an array section A_i representing the local index
set of the data array A on the I/O node i and $\bar{\mathbf{I}}^{D_i}$ denote an extended index
domain of a disk array section D_i representing the local index set of the disk
array D on the same I/O node. A distribution function of the array section
A_i with respect to the disk section D_i is defined by the mapping:*

$$\zeta_{D_i}^{A_i} : \bar{\mathbf{I}}^{A_i} \rightarrow \mathcal{P}(\bar{\mathbf{I}}^{D_i})$$

For each $i \in \mathbf{I}^A$, there are the following relations between the above introduced
mappings:

$$\gamma_D^A(\mathbf{i}) := \bigcup_{\mathbf{j} \in \beta_M^A(\mathbf{i})} \left(\bigcup_{\mathbf{i} \in \bar{\mathbf{I}}^{A_\mathbf{j}}} \zeta_{D_\mathbf{j}}^{A_\mathbf{j}}(\mathbf{i}) \right)$$

$$\beta_M^A(\mathbf{i}) := \{\mathbf{j} \mid \gamma_D^A(\mathbf{i}) = \mathbf{k} \wedge \mathbf{k} \in \bar{\mathbf{I}}^{D_\mathbf{j}}\}$$

Examples of file mapping will be introduced in Section 3.3.4.

3.3.4 Opening a Parallel File

Specification of the File Type. The OPEN statement establishes a connection between a unit[5] and an external file, and determines the connection properties by initializing specifiers in a *control list*. To be able to specify that the file is a parallel file or a standard one, the control list of the standard Fortran OPEN statement is extended by a new optional specifier FILETYPE. The value of the FILETYPE specifier must be *ST* when the file (standard file) is intended to be used for communication with the environment, or *PAR* when a parallel file is opened. The default is *ST*.

The meaning of these values is obvious from the examples below.

```
OPEN (8, FILE = '/usr/exa1', FILETYPE = 'ST', STATUS = 'NEW')
OPEN (9, FILE = '/usr/exa2', STATUS = 'OLD')
OPEN (10, FILE = '/usr/exa3', FILETYPE = 'PAR', STATUS = 'NEW')
```

Due to the above specifications, units 8 and 9 will refer to standard files (FILETYPE = 'ST'), and unit 10 will be connected to a parallel file (FILETYPE = 'PAR').

Remark: The OPEN statement also includes the FORM specifier which determines whether the file is being connected for formatted or unformatted I/O. We do not introduce this specifier in our examples because all the extensions discussed can be applied to both formatted and unformatted cases.

Array Section Access. Fortran supports the *sequential* file organization (the file contains a sequence of records, one after the other) and file organization known as *direct access* (all the records have the same length, each record is identified by an index number), and it is possible to write, read, or re-write any specified record without regard to position. It is possible to access an array element corresponding to a specific index value. However, the programmer has to handle the computation of the appropriate index number

---

[5] I/O statements do not refer to a file directly, but refer to a unit number, which must be connected to a file.

which is a little awkward in case of multi-dimensional arrays, and dependent on the storage organization supported by the given language system. A direct access to multi-dimensional array sections may be complicated.

Therefore, we introduce an additional type of the file organization referred to as **array section access** which is specified by assigning the value *ASA* to the standard specifier ACCESS. Such files are referred to as *array files*. This type of access is especially valuable for files storing whole arrays in the cases where providing efficient access according to the index or section specification is needed[6].

It is additionally necessary to specify the element size and shape of the array whose elements will be stored in the file. Therefore, we introduce a new specifier ARRAY_TYPE which has the form

$$ARRAY\_TYPE = (\text{array-element-type, array-shape})$$

The meaning of the specifier values is obvious from the examples below.

OPEN (11, FILE='/usr/exa4', FILETYPE='PAR', ACCESS='ASA', &
       ARRAY_TYPE =(REAL,(/50,100,200/)), STATUS ='NEW')

OPEN (12, FILE='/usr/exa5', FILETYPE='PAR', ACCESS='ASA', &
       ARRAY_TYPE =(REAL,SHAPE(B)), STATUS ='NEW')

OPEN (13, FILE='/usr/exa6', FILETYPE='PAR', ACCESS='ASA', &
       ARRAY_TYPE =ATYPE(A), STATUS ='NEW')

Due to the above specifications,

- unit 11 will refer to an array section access parallel file (FILETYPE = 'PAR' and ACCESS = 'ASA'). The file values will be viewed as real type elements of a three-dimensional array with shape (50,100,200).
- the shape of an array stored in the file connected to unit 12 is determined by the shape of the array *B*, and
- the element type and shape of the array stored in the file connected to unit 13 is determined by the (new) intrinsic function *ATYPE* which extracts the element type and shape from the argument array, array *A* in our example.

Explicit File Mapping Specification. The file that is being opened is bound to the appropriate FILEGRID specification that introduces a name that will denote the file, the dimensionality of the filegrid, and the specification of the distribution of the filegrid onto I/O nodes or disks.

---

[6] This functionality is important if the Vienna-Fortran/HPF programmer wants to explicitly program out-of-core applications.

```
MACHINE 'PARAGON (KFA JUELICH)'
PROCESSORS P1D(4)
IONODES IOP1D(2)
DISKS, DISTRIBUTED (BLOCK) TO IOP1D :: D1D(8)
FILEGRID, DISTRIBUTED (BLOCK) TO D1D :: F1(1000)
REAL, DISTRIBUTED (BLOCK) TO P1D :: A(1000)
OPEN (iu1, FILE = '/usr/exa4', FGRID = F1, ACCESS = 'ASA', &
        FILETYPE = 'PAR', ARRAY_TYPE = (REAL, (/100,200/))

WRITE (iu1) A
```

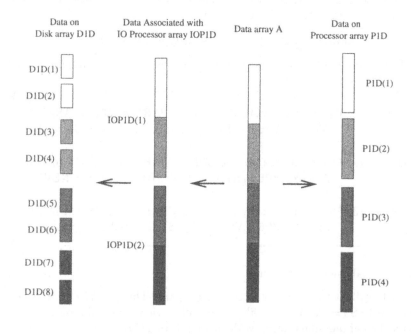

Fig. 3.15. Explicit File Mapping: Disks are Associated Blockwise with I/O Nodes.

Examples of the explicit specification of file mapping are introduced in Fig. 3.15 and 3.16. In the example introduced in Fig. 3.15, array A is distributed blockwise onto internal memory associated with four processors and onto eight disks, which are regularly distributed across two I/O nodes. Distribution of data onto disks and association of disks with I/O nodes indirectly determines the sets of indices of array elements, which are maintained by the appropriate I/O node. By means of the optional directive **MACHINE**, the user can give the compiler information about the target architecture. The compiler checks the corresponding configuration file to find out whether the

```
! No machine is specified; checking at runtime only
INTEGER, PARAMETER :: MAP(10) = (/1,4,4,3,1,1,2,4,3,4/)
PROCESSORS  P1D(4)
IONODES  IOP1D(4)
FILEGRID, DISTRIBUTED (INDIRECT(MAP),:) TO IOP1D :: F3(10,*)
REAL, DISTRIBUTED (INDIRECT(MAP),:) TO P1D :: E(10,2), G(10,3)
OPEN (iu3, FILE = '/usr/exa4', FILETYPE = 'PAR', FGRID = F3, &
            STATUS = 'NEW')
WRITE (iu3) E, G
```

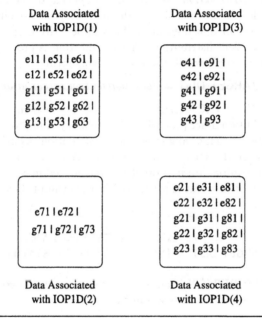

Fig. 3.16. Explicit File Mapping: The File Stores 2 Irregularly Distributed Arrays

specified resources, disks and I/O nodes are available. The strategy used, if the resources are not available is implementation dependent.

The code fragment in Fig. 3.16 illustrates explicit file mapping in the case when the file may have an unknown length (expressed by '*' in the definition of *F3*). Values of *MAP(1:10)* determine the distribution of the elements *F3(1:10,1)*, *F3(1:10,2)*, etc. The figure also shows the mapping of the array elements to the I/O nodes after the two irregularly distributed arrays *E* and *G* have been written to the file. In each list of data elements associated with an I/O node, two neighboring data elements are separated by '|'.

Specification of Hints. The main use of hints is, when the file is created, or when a distributed array is about to be written to a file, to tell the compiler/runtime system the intended use of the file in the future, so that it can organize the file data in an "optimum" way on the I/O system. The information that can be supplied by the programmer includes file access patterns hints, file data prefetching hints, data distribution hints, and other file specifics that can direct optimization. Hints do not change the semantics of parallel I/O operations. The language implementation is free to ignore some or even all hints. Below, the semantics of some hints is informally specified.

Target Internal Distribution. When a file is opened, or data is about to be written to a file (see Section 3.3.5), the user may pass a hint to the compiler/runtime system that data in the file will be written or read to/from an array of the specified distribution. The specification is provided by the hint specifier TARGET_DIST which has the following form

$$\text{TARGET\_DIST} = data\_distribution\_specification$$

Processing the TARGET_DIST specification results in the construction of a data distribution descriptor which, for an array A in processor p includes information about: shape, alignment and distribution, associated processor array, and size of the local data segment of A in each processor p (data distribution descriptors are in more detail introduced in Section 4.3).

Introduction of this hint is illustrated in the example below.

```
OPEN (20, FILE = '/usr/exa20', FILETYPE = 'PAR', &
      TARGET_DIST = DISTRIBUTION(A), STATUS = 'NEW')
```

where DISTRIBUTION is an intrinsic function which extracts
the distribution of its array argument

Each array will be written to the file *'/usr/exa20'* so as to optimize its reading into arrays which have the same distribution as array A.

In subsection 3.2.8, we briefly introduced user-defined distribution functions (UDDFs). The target internal distribution hint which refers a UDDF having the name *UserDef* and a list of arguments denoted by *arguments*, can have the following form:

$$\text{TARGET\_DIST} = \text{UserDef(arguments)}$$

The example below introduces a hint which refers a UDDF that specifies the regular distribution.

▼

```
! User-defined distribution function for regular distribution

    DFUNCTION  REG (M,N1,N2,K1,K2)
        TARGET_PROCESSORS  :: P2D(M,M)
        TARGET_ARRAY, ELM_TYPE (REAL) :: A(N1,N2)
        A  DISTRIBUTED ( CYCLIC(K1), CYCLIC(K2)) TO  P2D
    END DFUNCTION  REG

    OPEN (11, FILE ='/usr/exa3', FILETYPE ='PAR', STATUS ='NEW', &
                TARGET_DIST =REG(8,400,100,4,2))
```

▲

Here, by default, elements of all arrays will be written to the file *'/usr/exa3'* so as to optimize reading them into real arrays which have the shape (400,100), and they are distributed as (CYCLIC(4), CYCLIC(2)) onto a grid of processors having the shape (8,8). This predefined global I/O distribution specification can be temporarily changed by a **WRITE** statement (see Section 3.3.5).

In this way it is possible to introduce definitions of irregular and sparse matrix distributions, as illustrated below.

▼

```
! User-defined distribution function for irregular distribution

    DFUNCTION  IRREG (MAP,N)
        TARGET_PROCESSORS  :: P2D(*)
        TARGET_ARRAY, ELM_TYPE (REAL) :: A(N,10)
        INTEGER :: MAP(N)
        A  DISTRIBUTED ( INDIRECT (MAP),:) TO  P2D
    END DFUNCTION  IRREG

    OPEN (12, FILE ='/usr/exa4', FILETYPE ='PAR', STATUS ='OLD', &
                TARGET_DIST =IRREG(MAP1,N))
```

▲

Here, elements of all arrays will be written to the file *'/usr/exa4'* so as to optimize reading them into arrays which are distributed irregularly by the mapping array *MAP1* having *N* elements. The shape of the processor array will be determined from the context of the **OPEN** statements.

! User-defined distribution function for sparse matrix distribution

```
DFUNCTION  SPARSE_CRS_DIST (M,N,AD,AC,AR)
    TARGET_PROCESSORS :: P2D(0:M-1,0:M-1)
    TARGET_ARRAY , ELM_TYPE (DOUBLE PRECISION) :: A(N,N)
    !representation vectors
    INTEGER , DIMENSION (:) :: AD,AC,AR
    !sparse matrix distribution
    A DISTRIBUTED ( SPARSE ( CRS (AD,AC,AR))) TO P2D
END DFUNCTION  SPARSE_CRS_DIST

OPEN (13, FILE ='/usr/exa5', FILETYPE ='PAR', STATUS ='OLD', &
            TARGET_DIST =SPARSE_CRS_DIST(4,1000,AD,AC,AR))
```

Here, by default, elements of all arrays will be written to the file */usr/exa5* so as to optimize reading them into double precision sparse matrices which have the shape $(1000, 1000)$, and they are distributed due to the sparse matrix data distribution specification onto a grid of processors having the shape $(4, 4)$. The target arrays will be represented by three vectors, AD, AC and AR, using the CRS representation method (see Subsection 3.2.9).

For all possible STATUS options of the OPEN statement, the TARGET_DIST specifier introduces a very important hint specification. Data can be stored in the file in such a way that the highest data transfer bandwidth is assumed if the data is read into an array having the distribution specified by the TARGET_DIST specifier. When an old file is opened, it may automatically be reorganized if its data layout does not correspond to the one introduced by TARGET_DIST specification.

If a list of objects will be transferred to/from the file in one I/O operation, it is possible to specify for them a common distribution hint of the following form:

$$\text{TARGET\_DIST} = (\text{data\_distribution\_specification}_1 , \ldots ,$$
$$\text{data\_distribution\_specification}_n)$$

The use of this form of hint is illustrated by the example below.

```
OPEN (14, FILE ='/usr/exa6', FILETYPE ='PAR', STATUS ='OLD', &
        TARGET_DIST = (DISTRIBUTION(A), REG(8,400,100,4,2))
```

Here, pairs of arrays will be written to the file '/usr/exa6' so as to optimize reading them into pairs of arrays, in which the first array has the same distribution as array A and the distribution of the second array is determined by the UDDF specified on page 54.

File Access Hints. There are a lot of possible hint specifications which pass information about possible accesses to files to the compiler and runtime system. Specification of this set of hints is provided by the derived data type T_FAHINTS having the definition:

```
TYPE T_FAHINTS       ! derived type for file access hint specification
    INTEGER              :: PROCFILE   != NUMBER_OF_PROCESSORS()
    CHARACTER (LEN=5)  :: SECTION    != 'REG'
    CHARACTER (LEN=3)  :: SHARED     != 'NO'
    CHARACTER (LEN=5)  :: GRAN       != 'SEC'
    CHARACTER (LEN=5)  :: SYNCH      != 'SYNC'
    CHARACTER (LEN=3)  :: ORD        != 'YES'
    CHARACTER (LEN=3)  :: PREFETCH != 'NO'
    TYPE (MPIOT_HINTS) :: MPIOH
END TYPE T_FAHINTS
```

Variables of T_FAHINTS type are initialized by the procedure FAHINTS_INIT. The initial values are introduced as comments in the above definition of T_FAHINTS.

The specification of file access hints is provided by a new optional specifier FAHINTS in the OPEN statement. A statement sequence that will associate the specifier with file access hints is

```
TYPE (T_FAHINTS) :: FAH

CALL FAHINTS_INIT (FAH)

OPEN (u, FILE ='/usr/example', FILETYPE ='PAR', &
        TARGET_DIST =DISTRIBUTION(A), FAHINTS =FAH)
```

In the following, we introduce the semantics of the components that built-up the derived type T_FAHINTS.

The Number of Computational Nodes Participating in I/O. In an earlier study [37], Bordawekar et al. observed that the best performance need not necessarily be obtained when all processors performing computations also directly participate in I/O. With the hint PROCFILE, the user specifies the number of computational processors that should participate in performing I/O. By default, this component is initialized by the HPF intrinsic NUM-BER_OF_PROCESSORS.

Regularity of Accessed Sections. For the ASA file organization, it is useful to know whether regular sections, irregular sections or both kinds of them will be exchanged between the file and memory. This is specified by the component SECTION which can have 3 values: *REG, IRREG* and *MIXED*. A statement sequence that will specify this hint is

```
TYPE (T_FAHINTS) :: FAH
CALL FAHINTS_INIT (FAH)
FAH%SECTION = 'IRREG'
OPEN (u, FILE = '/usr/example', FILETYPE = 'PAR', FAHINTS = FAH)
```

Sharing Array Sections Among Processes. The component SHARED specifies in the form of 'YES' or 'NO' values whether during one I/O operation a file section will be shared among processes.

Granularity of Accesses. The component GRAN specifies whether the exchange of data with the file will be elementwise (value *ELM*), by sections (*SEC*), by whole arrays (*ARR*) or combined (*MIXED*).

Synchronous or Asynchronous Mode of I/O Operations. The component SYNCH specifies whether the file will be accessed exclusively synchronously (value *SYNC*), asynchronously (*ASYNC*), or using both modes (*MIXED*).

Data Ordering. In some applications the order of the data in the file does not matter. If the compiler has this information it can utilize it when looking for possibilities to run I/O concurrently with other I/O and for other optimizations. The user can declare a new component, ORD = *ord*, to determine whether the order of data is relevant or not; *ord* is a string having 'YES' or 'NO' values.

Data Prefetching Hints. When a file is being opened and PREFETCH = 'NO' is specified, no prefetching is applied (suggested) to this file in the scope of the whole program.

Hints Related to MPI-IO Hints. The component MPIOH denotes a set of hints that are derived from the file hints proposed by the MPI-IO Committee [87]. These hints are mainly concerned with layout of data on parallel I/O devices and access patterns. In this case, the hint specification is provided at low abstraction level.

3.3.5 Write and Read Operations on Parallel Files

All parallel files may be accessed via standard Fortran I/O statements that are implemented by parallel I/O operations wherever possible. Moreover, the language provides additional specifications for write statements that can be used by the programming environment to determine the sequence of array elements on the external storage and to optimize parallel accesses to them.

Writing to a File. The WRITE statement can be used to write multiple distributed arrays to a file in a single statement. Vienna Fortran provides several forms of the WRITE statement.

(i) In the simplest form, the individual distributions of the arrays determine the sequence of array elements written out to the file. For example, in the following statement:

$$\text{WRITE } (f, \text{ FILETYPE } = \text{'PAR'}) \; A_1, A_2, ..., A_r$$

where f denotes the I/O unit number connected to a parallel file by an OPEN statement, and A_i, $1 \leq i \leq r$ are array identifiers. This form should be used when the data is going to be read into arrays with the "same" distribution as A_i. This is the most efficient form of writing out a distributed array since each processor can independently (and in parallel) write out the part of the array that it owns, thus utilizing the I/O capacity of the architecture to its fullest.

(ii) Consider the situation in which the data is to be read several times into an array B, where the distribution of B is different from that of the array being written out. In this case, the user may wish to optimize the sequence of data elements in the file according to the distribution of the array B so as to make the multiple read operations more efficient. The TARGET_DIST specifier of the WRITE statement enables the user to specify the distribution of the target array in the same way that we introduced when we were discussing the OPEN statement. This additional specification can then be used by the programming environment to determine the sequence of elements in the output file.

$$\text{WRITE } (f, \text{ TARGET\_DIST } = \text{REG}(4,100,100,1,5)) \; A$$

Here, due to the specification of the distribution function REG (see page 54), the elements of the array A are written so as to optimize reading them into an array which has the shape (100,100) and is distributed as (CYCLIC *(1)*, CYCLIC *(5)*) onto a grid of processors having the shape (4,4).

The target data distribution hint of the WRITE statement can also have a simplified form:

TARGET_DIST = *(distribution-format-list)*

as illustrated by the following example

WRITE (f, TARGET_DIST = (BLOCK (M), CYCLIC)) A

Here, the complete data distribution specification is constructed from the (BLOCK (M), CYCLIC) annotation, the shape and type of A, and the processor array which is the target of the distribution of A.

(iii) If the data in a file is to be subsequently read into arrays with different distributions or there is no user information available about the distribution of the target arrays, automatic processing mode is applied. The implementation of this feature includes automatic derivation of hints for optimization of the data layout on disks, transformations to insert the necessary reorganization of data on disks, improve locality, and overlap I/O with computation and other I/O.

Other examples of writing data to a parallel file are introduced below.

! The following two write statements have the same effect on the
! target file:
! they write the data in the file position reserved for the element of
! a two-dimensional array having index (15,40)
WRITE (11, SECTION = (15,40)) k
WRITE (11, SECTION =(15,40)) a(15,40)

! 2000 elements of C are written into the file positions reserved
! for the array section (1:100, 31:50)
WRITE (u, SECTION = (1:100, 31:50)) C(1:2000)

! section (201:400, 1:20) of D is written into the file positions reserved
! for the array section (1:200, 31:50)
WRITE (u, SECTION = (1:200, 31:50)) D(201:400, 1:20)

Remark: If the TARGET_DIST or SECTION specifiers are introduced in the WRITE operation, the file is considered as parallel by default.

Reading from a File. A read operation to one or more distributed arrays is specified by a statement of the following form:

READ (f, FILETYPE = 'PAR') B_1, B_2, ..., B_r

where again f denotes the I/O unit number and B_i, $1 \leq i \leq r$ are array identifiers.

3.3.6 I/O Alignment

So far, we have specified the external distribution of an array explicitly either by means of a filegrid object or by hints. Another way to specify an external array's distribution is by referring to the external distribution of another, previously distributed array, which is called the *source* array.

By the following specification

```
REAL, DISTRIBUTED (BLOCK) :: A(1000000), B(500000)
B(:) IOALIGN WITH A(500001:1000000)
```

the user gives the programming environment an advice to store the i-th element B on the same disk as the $(500000+i)$-th element of A. This relation should hold by any reorganization of the file which stores A or prefetching copies of the elements $A(500001 : 1000000)$ to the "nearest" disks.

As in the case with alignment in internal memory, the user may define functions to describe more complex alignments.

3.3.7 Other I/O Statements

Reorganizing a Parallel File. The REORGANIZE statement enables the user to specify restructuring a file into a form which will be more efficient for the subsequent treatment.

Transfer of Data Between Standard and Parallel Files. The statement COPYIN copies standard files into parallel files and the statement COPYOUT copies parallel files onto standard files (see Fig. 3.12).

Asynchronous I/O. The WRITE, READ, REORGANIZE, COPYIN and COPY-OUT statements may include an EVENT specifier to specify that the I/O operation is to take place asynchronously (that is, while other processing continues). This specifier introduces an event variable. The waiting on an I/O event is specified by the WAIT statement which takes the event variable as an argument.

3.3.8 Intrinsic Procedures

An important feature of HPF and Vienna Fortran is a variety of intrinsic functions and subroutines to support efficient programming. The intrinsic procedures related to parallel I/O are briefly summarized below.

- System inquiry functions. They return values that describe the size and shape of the underlying disk and I/O node arrays and the total number of disks connected to the I/O node passed as an argument.
- Mapping inquiry procedures. They allow the program to determine the actual mapping of an array, an array section or an array element with regard to a disk or I/O node at runtime.
- Hints accessing procedures. They allow the program to inquire or change all or individual hints.

3.3.9 Experiments

In this subsection, we present some performance measurements to justify the need for user control over the manner in which data from distributed arrays is transferred to and from secondary storage. We consider parallel I/O of regularly distributed arrays, irregularly distributed arrays, and distributed sparse matrices.

Parallel I/O of Regularly Distributed Arrays

Consider the following declarations:

```
PARAMETER :: NP =..., N =...
PROCESSORS :: P(NP, NP)
REAL, DISTRIBUTED ( BLOCK, BLOCK )  TO  P :: A(N,N)
```

Here, A, is an $N \times N$ array, block distributed in both dimensions across an $NP \times NP$ processor array. Fig. 3.17 shows the distribution of elements of the array A for the case of $N = 4$ and $NP = 2$.

	A(1,1) A(1,2)	A(1,3) A(1,4)	
P(1,1)			P(1,2)
	A(2,1) A(2,2)	A(2,3) A(2,4)	
	A(3,1) A(3,2)	A(3,3) A(3,4)	
P(2,1)			P(2,2)
	A(4,1) A(4,2)	A(4,3) A(4,4)	

Fig. 3.17. A Two-Dimensional Block Distributed Array

If such an array is written out using a standard FORTRAN WRITE statement, the semantics enforce the column-major linearization of the data elements. This would require close synchronization of the processors owning A to execute the WRITE statement. Besides this serialization, another drawback is that each processors writes only small blocks of the individual columns.

On most systems, such as the iPSC/860 Concurrent File System (CFS), the best performance for I/O operations is reached for large blocks. The same inefficiencies recur, if the data has to be subsequently read into a similarly block distributed two-dimensional array.

On the other hand, if we enable each process to write its local elements as one block, in parallel with other processes, the sequence of the data elements in the file would be as follows:

A(1,1), A(2,1), A(1,2), A(2,2), A(3,1), A(4,1), A(3,2),
A(4,2), A(1,3), A(2,3),A(1,4), A(2,4), A(3,3),
A(4,3), A(3,4), A(4,4)

Similarly, reading the data into a similarly distributed array can also be executed in parallel.

In order to determine the overheads involved in writing out an array distributed as described above, we implemented five versions of the write statement on the Intel iPSC/860. The system consists of 32 processing nodes and 4 I/O nodes using CFS to manage the file system.

The first four versions of our experiment, preserve the standard FORTRAN linearization order, while the last uses the sequence suggested above.

In the first implementation, *CENT*, each process sends its local block of elements to a designated process which collects the entire array. This central process then writes the array out to the CFS using a standard FORTRAN write statement. The next three implementations, *SEQ0*, *SEQ1* and *SEQ2* again preserve the column-major linearization of the array and use CFS's file modes 0, 1 and 2 respectively [89], to write out the array.

In *SEQ0*, each process manages its own file pointer. All processes write unsynchronized to the same file. They position their file pointer to the appropriate position in the file for each subcolumn that they have to output.

The processes work with a common file pointer in version *SEQ1* and thus have to be closely synchronized. For each part of a column, the appropriate process performs the write while the other processes are waiting.

In *SEQ2*, the write operations are executed as collective operations. The columns are written sequentially. Thus, each process which owns a part of the column writes its part. Other processes perform the write with zero length information. The information written in such a collective operation is ordered in the output file according to the process numbers.

The last version, *NEW*, uses the experimental implementation in which, instead of writing out the data in the column-major order, each process writes out its local piece as contiguous block. The processes perform a single collective write using the CFS's file mode 3.

Table 3.1 shows the times measured for a 1000 × 1000 array distributed blockwise across a 4 × 4 processor array. Since the performance depends heavily on whether the file to be written exists prior to the operation or

Version	Including file creation	Pre-existing=1B file
CENT	4.3	4.0
SEQ0	52.2	6.5
SEQ1	43.9	7.9
SEQ2	42.3	4.4
NEW	1.9	1.6

Table **3.1.** Time (in secs) for Writing out a Distributed Array

not, we present timings for both cases. The problem is that if the file does not already exist, new disk blocks have to be allocated every time the file is extended. This is particularly an issue with the versions, *SEQ0, SEQ1* and *SEQ2* since each individual write for a part of the column extends the file.

It is clear from our experiments, that at least on the iPSC/860, that the version *NEW* performs better than the rest of the implementations. This indicates that I/O bound applications running on distributed memory machine may achieve much better performance if the user can provide information which would help the compiler and runtime system to choose the best possible sequence of the data elements written out to secondary storage.

Parallel I/O of Irregularly Distributed Arrays

Consider the following declarations:

```
PARAMETER :: NP = ..., N = ...
PROCESSORS :: P(NP)
REAL , DYNAMIC :: A(N), B(N)
INTEGER , DISTRIBUTED (BLOCK) :: MAP1(N), MAP2(N)
... initialization of MAP1 and MAP2 ...
DISTRIBUTE A :: ( INDIRECT (MAP1))
DISTRIBUTE B :: ( INDIRECT (MAP2))
```

Here, *A* and *B*, are indirectly distributed across a one-dimensional processor array.

In order to determine the overheads involved in writing out array *A* and reading in arrays with different distributions, we implemented four versions of the WRITE statement and three versions of the READ statement on the Intel Paragon XP/S10, installed at KFA Juelich, Germany, using the Parallel File System (PFS) [105] to manage parallel files. The system consists of 138 compute nodes and the PFS configuration consists of five RAID systems having five disks each.

The features of all the implementations, denoted by *write1, write2, write3, write4, read1, read2,* and *read3*, can be specified as follows.

write1: When transferring elements of array A to a parallel file, each node does an output operation controlling the transfer of local part of the array to the corresponding part of the file. Moreover, the distribution descriptor of A is appended as metadata to the file. This implementation is suitable for cases in which the data written will be dominantly read into arrays having the same distribution as A, but may also be read into an array X, for example, where the distribution of X is different from that of array A.

write2: Here, the elements of the array A are written so as to optimize reading them into array B. A *two-phase access strategy* [37, 99] is used. In the first phase, the data is copied into a temporary array T having the same distribution as B using procedures from the CHAOS runtime library [217]. In phase 2, each node does an output operation controlling the transfer of local part of array T to the corresponding part of the file.

write3: The two-phase access strategy is used as in the *write2* implementation. However, the distribution descriptor of B is appended to the file in this case.

write4: The elements of A are written to a parallel file in a standard Fortran sequence using the two-phase access strategy. In the first phase, the data is copied into a temporary array T having the same size as A and being distributed by BLOCK. In phase 2, each node does an output operation controlling the transfer of local part of array T to the corresponding part of the file.

read1: When transferring the elements of A from a parallel file, this implementation assumes that the elements are ordered according to the standard Fortran sequence. The two-phase strategy is applied; the elements are first read into a temporary array T having the block distribution and then remapped into array A using the CHAOS library.

read2: This implementation is complementary to *write1*. The layout of data elements in a parallel file is optimized for the reading into an array having the same distribution as A. Reading into B is implemented using the two-phase strategy .

read3: The data in the PFS file were prepared for the reading onto B by *write2*.

We have obtained performance evaluation results for various experiments. The execution times were measured for different numbers of computational nodes and different array sizes. In these experiments, we were concerned not with the total time required to process a certain number of bytes with a READ or WRITE operation, but rather with the number of bytes that can be processed per second - *throughput*.

write	array size: 5 MBytes		
nodes	32	64	128
write1	1.40	0.75	0.38
write2	0.25	0.41	0.33
write3	0.23	0.31	0.22
write4	0.20	0.31	0.34

Table 3.2. WRITE Implementations. Throughput in MBytes/sec

read	array size: 5 MBytes		
nodes	32	64	128
read1	2.12	1.78	1.75
read2	0.31	0.97	0.53
read3	0.07	0.17	0.19
read4	2.86	1.52	0.76

Table 3.3. READ Implementations. Throughput in MBytes/sec

Tables 3.2 and 3.3 respectively present the performance of the WRITE and READ implementations when transferring the elements of a 5 MByte array. It is clear that the versions write1 and read1 perform better than the other implementations. If a file is written to from an array having a specific irregular distribution, and is read into an array having a different irregular distribution, the best performance is achieved by the combination write2/read3, as shown in Table 3.4. We also observed the same effect when experimenting with 0.25, 0.5 and 2 MByte arrays.

write + read		array size: 5 MBytes		
nodes		32	64	128
write1	read2	0.07	0.14	0.13
write2	read3	0.23	0.32	0.23
write3		0.22	0.26	0.17
write4	read1	0.12	0.24	0.21

Table 3.4. WRITE/READ Combinations. Throughput in MBytes/sec

In all the experiments, degradation in the performance was observed for more than a certain number of processors, due to a large synchronization overhead. Therefore we investigated the impact of the number of computational nodes executing parallel file operations on the performance of a READ or WRITE operation.

Tables 3.5 and 3.6 present the performance observed when either all computational processors that the array is distributed onto, or a subset of them participate directly in the parallel file operation. We used the CHAOS library to move (gather) the array elements into the processors which are going to write them into the file and to move (scatter) the array elements from the processors which have read them from the file.

Note that the programmer can use the PROCFILE hint to suggest the number of nodes to be used in executing the file operations.

write								array size: 2 MB	
nodes	16	32		64			128			
	all	all	1/2	all	1/2	1/4	all	1/2	1/4	1/8
write1	0.83	0.34	0.59	0.23	0.33	0.50	0.13	0.17	0.24	0.37
write2	0.17	0.32	0.37	0.27	0.39	0.51	0.17	0.24	0.32	0.39
write3	0.15	0.26	0.31	0.22	0.32	0.37	0.10	0.19	0.26	0.31
write4	0.15	0.31	0.41	0.32	0.57	0.58	0.19	0.38	0.37	0.43

Table 3.5. WRITE Implementations; the Number of the Compute Nodes Participating in I/O Varies. Throughput in MBytes/sec

read								array size: 2 MB	
nodes	16	32		64			128			
	all	all	1/2	all	1/2	1/4	all	1/2	1/4	1/8
read1	0.64	0.60	0.88	0.39	0.64	0.79	0.19	0.35	0.44	0.53
read2	0.14	0.28	0.30	0.25	0.29	0.46	0.13	0.24	0.32	0.40
read3	1.48	0.76	1.14	0.39	0.66	0.82	0.20	0.29	0.46	0.61

Table 3.6. READ Implementations; the Number of the Compute Nodes Participating in I/O Varies. Throughput in MBytes/sec

Parallel I/O of Distributed Sparse Matrices

Consider an integer sparse matrix M of shape (1000,1000) which includes 100000 non-zero elements. In our experiments, the matrix will get either MRD or CYCLIC distribution (see 3.2.9).

In order to investigate the overheads involved in writing out matrix M and reading in matrices with different distributions, we implemented several versions of the WRITE statement and the READ statement on the Intel Paragon XP/S10. The configuration of this system was described in the previous paragraph. On this system, it is is necessary to allocate 390 KBytes for the non-zero elements of the sparse matrix M.

The features of all the implementations can be specified as follows.

readndsr: s-representation → d-representation
The sparse matrix stored on disks in the CRS s-representation is read into the sparse matrix M having the CYCLIC or MRD distribution.

readdsr: d-representation → d-representation
The sparse matrix already stored on disks in the d-representation is read into the sparse matrix M having the CYCLIC or MRD distribution.

writendsr: d-representation → s-representation
The d-representation of the sparse matrix M having the CYCLIC or MRD distribution is transformed into the s-representation and M is written to disks in this new form.

writedsr: d-representation → d-representation
The sparse matrix M having the CYCLIC or MRD distribution is written on disks in its respective d-representation.

mrdtocyc: d-representation(mrd) → d-representation(cyclic)
The sparse matrix stored on disks in the d-representation of the MRD distribution is read into the sparse matrix M having the CYCLIC distribution.

cyctomrd: d-representation(cyclic) → d-representation(mrd)
The sparse matrix stored on disks in the d-representation of the CYCLIC distribution is read into the sparse matrix M having the MRD distribution.

(cyc/mrd)readdsrfewer: d-representation(cyclic/mrd) →
d-representation(cyclic/mrd)
The sparse matrix stored on disks in the d-representation of the CYCLIC (MRD) distribution onto p processors is read into the sparse matrix M having CYCLIC (MRD) distribution onto q processors, whereby $p > q$.

(cyc/mrd)readdsrmore: d-representation(cyclic/mrd) →
d-representation(cyclic/mrd)
The sparse matrix stored on disks in the d-representation of the CYCLIC (MRD) distribution onto p processors is read into the sparse matrix M having CYCLCIC (MRD) distribution onto q processors, whereby $p < q$.

Sparse Matrix (1000 columns, 1000 rows) size of data: 390 KB distribution type: CYCLIC			
nodes	4	16	64
readndsr	2.12	3.66	10.78
readdsr	0.49	2.84	6.66
writedsr	0.93	1.49	10.43
writendsr	6.36	2.96	8.76

Table 3.7. READ and WRITE Implementations for the CYCLIC Distribution. Time in Seconds

Sparse Matrix (1000 columns, 1000 rows) size of data: 390 KB distribution type: MRD			
nodes	4	16	64
readndsr	1.70	3.15	11.36
readdsr	0.53	1.61	5.85
writedsr	0.88	3.75	9.67
writendsr	3.17	2.73	7.92

Table 3.8. READ and WRITE Implementations for the MRD Distribution. Time in Seconds

Tables 3.7 and 3.8 show that the versions *readdsr* and *writedsr* perform for both types of distribution better than *readndsr* and *writendsr*; the improvement is especially remarkable when 64 processors are used.

For the given sparse matrix size and initialization, the best results for the MRD / CYCLIC and CYCLIC / MRD redistribution were achieved when 16 nodes were used (Table 3.9).

Sparse Matrix (1000 columns, 1000 rows) size of data: 390 KB			
nodes	4	16	64
mrdtocyc	5.76	2.59	7.86
cyctomrd	2.86	2.66	7.21

Table 3.9. READ Implementations with MRD to CYCLIC and CYCLIC to MRD Redistributions. Time in Seconds

Sparse Matrix (1000 columns, 1000 rows) size of data: 390 KB			
nodes used: before/now	16/4	64/4	64/16
mrdreaddsrfewer	3.31	3.32	2.69
cycreaddsrfewer	8.51	5.47	2.92

Table 3.10. Reading into a Smaller Number of Processors. Time in Seconds

Sparse Matrix (1000 columns, 1000 rows) size of data: 390 KB			
nodes used: before/now	4/16	4/64	16/64
mrdreaddsrmore	1.74	1.76	2.83
cycreaddsrmore	1.82	1.83	2.84

Table 3.11. Reading into a Greater Number of Processors. Time in Seconds

Tables 3.10 and 3.11 show that even if the reading needs the redistribution of data, the performance is better in comparison with the case when the matrix is stored in the s-representation.

The performance results presented indicate that an I/O bound sparse matrix application may achieve much better performance if the user can provide information which would help the compiler and runtime system to choose the best organization of the data elements written out to secondary storage.

3.4 Out-of-Core Annotation

As we already mentioned in Section 1.1, many potential applications of high performance languages operate on large quantities of data. Primary data structures for these applications reside on disks. Therefore, these data structures are termed as out-of-core (OOC) data structures.

In order not to restrict application developers to problem sizes that fit in the memory of a system, providing a way to specify OOC data structures from Vienna Fortran is considered important. The mechanisms for providing this feature include extending the Vienna Fortran annotation to declare out-of-core arrays and their parts in main memory.

The Vienna Fortran OOC annotation is based on the OOC programming model which does not appear to the scientific application programmer very different from the in-core one[7]. The goal is to preserve for the programmer the *model of unlimited main memory*. It is assumed that the programmer writes an in-core version of the Vienna Fortran program which is then converted into the appropriate OOC Vienna Fortran form. According to the level of effort that must be provided by the programmer, we consider two modes in which the conversion can be done: (1) user controlled and (2) automatic. The modes are described in parts 3.4.1, and 3.4.2 in more detail.

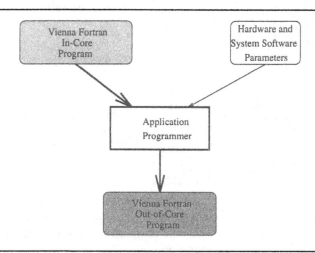

Fig. 3.18. Out-of-core Program Development

3.4.1 User Controlled Mode

In order to allocate memory and process accesses to OOC arrays, the programming environment needs information about which arrays are out-of-core and also the maximum amount of in-core memory that is allowed to be allocated for each array.

A list of arrays may obtain the out-of-core attribute in a type declaration statement such as

---

[7] Note that when developing in-core Vienna Fortran programs the programmer has only to specify data distribution and in some cases also work distribution.

PARAMETER :: M = ..., N = ...
REAL , DISTRIBUTED (BLOCK), OUT_OF_CORE , IN_MEM (M)::A(N),B(N)

Here, arrays A and B are out-of-core and if the machine nodes had infinite memory the elements of these arrays will be distributed blockwise across this memory. The part of the declaration introduced by the keyword IN_MEM indicates that at most M elements of each OOC array specified by the given declaration may be held in memory. The IN_MEM specification is optional.

A graphical sketch of how the in-core/out-of-core conversion can be done is depicted in Fig. 3.18. When applying this mode, the programmer analyzes the Vienna Fortran program and predicts memory requirements of the program after its parallelization. If there might not be enough main memory for the in-core (IC) implementation on the given target architecture, the programmer annotates some large data structures that have to be processed using OOC techniques (see 6.2). The programmer's decision is also based on her or his knowledge of features of the target system hardware (memory capacity) and software (memory requirements).

All computations are performed on the data in processors' internal memory. VFCS restructures the source out-of-core program in such a way that during the course of the computation, sections of the array are fetched from disks into the internal memory, the new values are computed and the sections updated are stored back onto disks if necessary. The computation is performed in stages where each stage operates on a different part of the array called a *slab* (or a tile). Loop iterations are partitioned so that data of slab size can be operated on in each stage. Each processor's local memory sees the individual slabs through a "window" referred to as the in-core portion of the array. VFCS has to get the information which arrays are out-of-core and what is the shape and size of the corresponding in-core portions in the form of OOC annotation.

Formally, the *OOC array annotation* is of the following form:

$$atype, dist\_spec, \text{OUT\_OF\_CORE } [, \text{IN\_MEM } (ic\_portion\_spec)] :: ad_1, .., ad_r$$

where *atype* specifies the type of the array elements, ad_i, $1 \leq i \leq r$ specify array identifiers B_i and their index domains, and *dist_spec* represents a Vienna Fortran *distribution-specification* annotation. The keyword OUT_OF_CORE indicates that all B_i are out-of-core arrays. In the optional part, the keyword IN_MEM indicates that only the array portions of the shape that corresponds to *ic_portion_spec* are allowed to be kept in memory. The larger the IC portion the better, as it reduces the number of disk accesses.

The combined data distribution and out-of-core annotation, and relation of slabs and in-core portions are illustrated by the example in Fig. 3.19 where the out-of-core annotation specifies that each local segment of matrix A is partitioned columnwise into slabs of 250 elements.

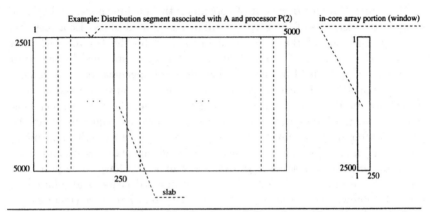

Fig. 3.19. Interaction of Data Distribution and Out-of-core Program Annotation

Another way for specifying the out-of-core attribute is the *out-of-core specification* statement. The following specification is equivalent to the one introduced on page 70.

REAL, DISTRIBUTED (BLOCK) :: A(N), B(N)
OUT_OF_CORE, IN_MEM (M) :: A, B

All arrays of a program unit may be specified as out-of-core in the following way:

OUT_OF_CORE, IN_MEM (M) :: ALL

One can specify a particular memory size for the processors (nodes) of the system by using

MEMORY Mb(n)

specification. This specifies that a system partition will be allocated all of whose nodes have at least n Megabytes of memory. Then it is not necessary to introduce the IN_MEM specification. The illustrative example follows.

PARAMETER :: N = ..., K = ...
MEMORY Mb(K)
REAL, DISTRIBUTED (BLOCK , BLOCK) :: A(N,N)
REAL, ALIGNED WITH A :: B(N,N), C(N,N)

OUT_OF_CORE :: A, B, C

_____▲

For the above example, the programming environment automatically derives the IN_MEM specification using the memory constrains specification introduced by the MEMORY construct. A possible approach is introduced in [155].

Restrictions: One cannot align an in-core array with an out-of-core array because it is difficult to enforce the meaning of align in such cases.

Association of Out-of-Core Arrays with Files. If the data for an out-of-core array comes from an input file then the opening and reading of the file, as it used in the in-core program version is replaced by a statement that associates the file with the out-of-core array. This is different from traditional open and read/write of a file because the user does not explicitly access the file.

We use the filegrid concept introduced in Section 3.3.3 to specify the association between file records and OOC arrays. The notation is based on the proposal of Choudhary et al. [77].

A file represented by a filegrid F is associated with an out-of-core array A by a statement

ASSOCIATE (F[, pos]) :: A

where *pos* denotes the optional specification of the starting position of the dataset of A within the file. By default, for example, the element $A(1,1)$ of a two-dimensional matrix corresponds to the first element of the file.

3.4.2 Automatic Mode

Scientific data-parallel programs tend to be enormous (thousands or ten thousands of lines). It is difficult for the user to annotate such programs for OOC arrays without knowing details of the compiler's parallelization strategy. The compiler may introduce additional temporaries and expands the Vienna Fortran code into an SPMD form in a way that is unpredictable for the user. Therefore, the resulting out-of-core program may still crash because of the memory lack. In this case the user has to update the current OOC Vienna Fortran program version to further reduce the memory requirements. On the other hand, if the user's decision was too conservative, the program runs are safe but the program performance may be superfluously lowered because of the significant overhead of the OOC implementation.

Only a few expert users are willing to invest a large effort into the analysis of program requirements on memory and to study the parallelization strategy of the restructuring compiler. The majority of the users prefers a compilation

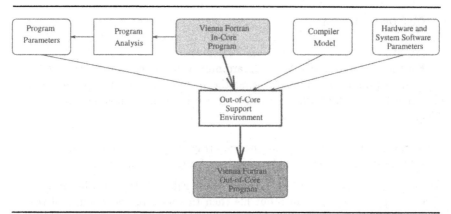

Fig. 3.20. Automatic Out-of-core Program Annotation

environment which automatically recognizes the need for the OOC implementation and essentially performs an automatic translation of an IC (in-core) Vienna Fortran program into the appropriate OOC form. The functionality of such a system is depicted in Fig. 3.20. It is able to predict the IC memory lack and to optimize the selection and annotation of OOC arrays with the aim to minimize the amount of I/O, to balance the I/O time with computation, and to fit the portions within physical memory. To make decisions, it utilizes the results of program analysis and the parameters of the compiler, hardware, and system software models. The program analysis identifies, for example, the run-time intensive program portions and estimates the buffer size for copies of non-local data. In some cases this information is provided in symbolic form. It is then the task of the user to interactively replace some symbolic expressions by approximate values. In many cases the compile time analysis can be combined with the runtime analysis.

BIBLIOGRAPHICAL NOTES

Some of the related research in language development for parallel I/O is briefly discussed below.

In recent years, only a small effort has been put into the design of language extensions supporting parallel I/O. Vienna Fortran was the first language which introduced I/O operations on distributed arrays. The proposal [55] introduces array files that may contain values from more than one distributed array. The array files are logically structured into records. Each record contains an array distribution descriptor (metadata) followed by a sequence of data elements associated with the array. The proposal also introduces new constructs to specify parallel I/O operations on array files. The parallel write operation enables the user to provide information about the distribution of the target array.

Marc Snir [236] has proposed two independent approaches how to expand the notation of HPF to include file operations for high performance I/O. One of them is based on the Vienna Fortran proposal which is outlined above. The second one provided us with inspiration for the introduction of the FI-LEGRID concept and annotation for mapping of arrays to disks and I/O nodes.

The Fortran 95 proposed draft standard X3J3/96-007r1 and the HPF-2 proposed draft [143] propose mechanisms for performing asynchronous I/O from an HPF or Fortran program. These introduce additional I/O control specifiers and the WAIT statement. Their approach is similar to the one that we presented in the context of Vienna Fortran.

Bordawekar and Choudhary [39] have proposed some directives for parallel I/O that can be used in conjunction with other HPF directives. In particular, they have introduced a directive for the specification of a logical disk array, an array of processors which really participate in performing I/O, a file template that is distributed across the disk array, and association of an array template with a file template. Regarding the abstraction level, the Bordawekar and Choudhary approach is comparable with the prescriptive model introduced in this paper.

MPP Fortran [207] treats I/O in a very limited extent. By a special directive it is possible to distinguish I/O operations on distributed and replicated variables.

Proposals for out-of-core annotation of HPF programs are introduced in [155, 43, 206]

The MPI-IO document [87] provides a means for passing file hints to the MPI-IO library interface.

A two-phase optimization strategy for parallel I/O of regularly distributed arrays has been proposed by Bordawekar et al. [37, 99].

4. Compiling In-Core Programs

In this chapter, we deal with the automatic parallelization of programs whose data structures may be completely stored in main memory; such programs are called *in-core programs*. The main focus is on irregular in-core programs. The reader will note that our use of the term *compiler* or *compilation system* will be rather general: in particular, we include interactive and automatic transformation systems in this concept.

4.1 Parallelizing Compilation Systems

The parallelization process as described in this chapter is based on the *SPMD* [156], model of computation (see part 2.2.1). Starting from the user specified data alignment and data distribution directives the compiler has to determine the layout of the data on the set of virtual processors. By default, work distribution is achieved by applying the *owner computes* rule, so that each processor only computes those data elements that are allocated in its local memory. However, in some cases this rule may be changed by the user or a special tool for automatic data and work distribution that is coupled to the compilation system. Access to non-local data is handled via explicit message passing; whenever access to non-local data is necessary communication constructs must be inserted by the compiler in order to send and receive data at the appropriate positions in the code.

The development of a message passing code can be done in two basic ways:

1. *Source-to-Source Solution*
   The parallelizing environment is built on top of an existing compiler as illustrated in Fig. 4.1 (where RTS indicates the runtime system). The source program is processed by the frontend and transformed into an internal representation (i.e. abstract syntax tree). The front end performs the analysis tasks of a conventional compiler, that is scanning, parsing and static semantic analysis. Among the other tasks performed by the front end are program analysis (including flow analysis and dependence analysis) and syntactic normalization. Normalization has the purpose of

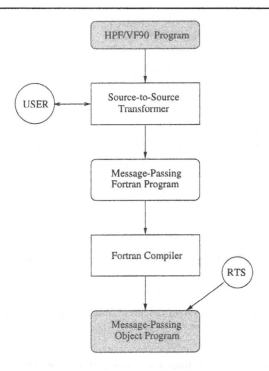

Fig. 4.1. The Compilation System Based on the Source-to-Source Parallelizer

generating a standard form of the program in order to simplify subsequent compiler optimizations. Normalization transformations typically include loop normalization, if conversion and subscript standardization. Array declarations, for example, are transformed in such a way that the lower bound in each dimension is 1, and all corresponding references are transformed accordingly. Subscript expressions are standardized in order to simplify data dependence analysis. Furthermore, standard transformations like constant propagation and dead code elimination are applied. For details, about these transformations see for example [24, 274]. The normalized internal representation of the HPF/VF90 program[1] is coverted into a message passing form in several parallelization phases, and then a message-passing Fortran program for a concrete target machine is generated from the internal program by the backend. Finally, the message-passing Fortran program is translated by the Fortran compiler into the object program.

---

[1] Remember that HPF/VF90 stands for High Performance Fortran/Vienna Fortran 90.

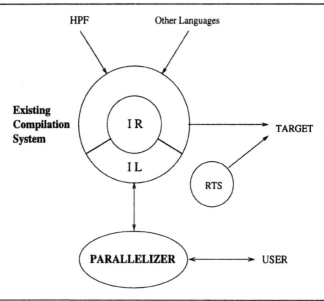

Fig. 4.2. The Compilation System Oriented Towards Direct Generation of Parallel Code

The application of transformations can result in greatly improved code if the user supplies the system with additional information. The user may change the specification of the distribution of the program's data via an interactive language. Program flow and dependence analysis information, using both intraprocedural and interprocedural analysis techniques, is computed and made available to the user, who may select individual transformation strategies or request other services via menus.
A great disadvantage of this approach is that it leads to a considerable duplication of effort, because the communication between the transformer and the Fortran compiler is one-way, i.e. the transformer cannot take advantage of the analysis that many modern compilers already incorporate. Further, it is difficult to predict the performance of the object code generated by the Fortran compiler for the target processor. The great advantage of this approach is that the high portability of the parallelized code can be achieved if a standard message passing interface, like MPI, is used for the implementation of communication.

2. *Source-to-Object Solution: An Integrated Environment Coupling a Parallelizer to a Compilation System*
In this approach, the parallelization part, which is referred to as *PARALLELIZER*, makes use of an existing compilation system originally built for translation of sequential languages by gaining access to its internal

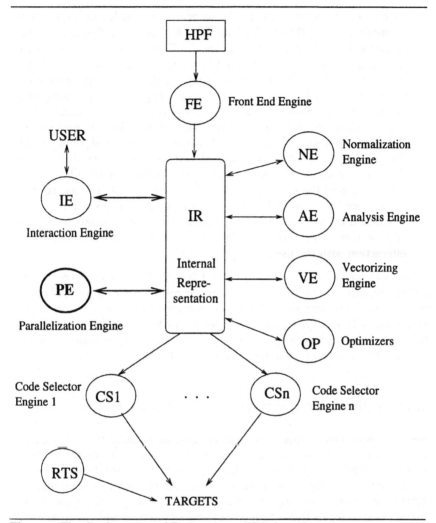

Fig. 4.3. The Configuration of Engines in an HPF Compiler

representation through the standardized Interface Layer (IL) (see Fig. 4.2). The facilities of the internal representation are enhanced in order to accommodate the information needed for parallelization. The *PARAL-LELIZER* benefits from the elaborate analysis and performance prediction that is performed by the compilation system. Recently, this approach was taken when developing the PREPARE HPF compiler [12, 252] which relies on an innovative compilation framework ([10]) that makes it possible to configure highly optimizing compilers from a large set of building blocks called *engines*. These engines work concurrently (when they run on a set of processors the compilation process is speeded up); they share a generic Internal Representation (IR)[2]. A possible configuration of the compiler is depicted in Fig. 4.3. The HPF program to be compiled is transformed into the IR by the HPF Front-End engine. Analysis engines enrich the IR with information necessary for the normalizations and optimizations. A dedicated engine, vectorizing engine, performs automatic DO loop vectorization. The *Parallelization Engine* (PE) transforms the normalized program IR into a form that can be efficiently executed on the target distributed-memory architecture. Some transformations are chosen on the basis of performance information provided by the Cost Estimator Module at compile time, or by the Evaluation Module during a previous run of the parallelized program. Then, the IR is translated into object code by a Code Selector; on the engine-like basis. The binary code generation focuses on gaining high performance by integration of intra- and interprocessor parallelism. The interactive engine (IE) controls all interaction with the user.

In both approaches, all analyses and transformations are done on the internal representation. In spite of this, in the rest of the book we shall ignore the internal representation and discuss programs and their transformations at the source level only, with the aim to make the text as readable as possible. However, the developed concepts may be fully utilized in both approaches to the parallelization and compilation discussed above.

4.2 Parallelization Strategies for Regular Codes

Regular codes can be precisely characterized at compile time. We distinguish between codes specifying *loosely synchronous computations* and codes specifying *pipelined computations*. Loosely synchronous problems ([115]) are divided into cycles by some parameter which can be used to synchronize the different processors working on the problem. This computational parameter corresponds to for instance, time in physical simulations and iteration count in the solution of systems of equations by relaxation. In such problems, it is only necessary to identify and insert efficient vector or collective communication at appropriate points in the program[3] (see Fig. 4.4(a)). The Jacobi code in Fig. 4.5 is a typical representative of this class.

A different class of computations contains *wavefront parallelism* because of existing loop-carried cross-processor data dependences that sequentialize computations over distributed array dimensions. To exploit this kind of parallelism, it is necessary to pipeline the computation and communication (see Fig. 4.4(b)).

---

[2] In this IR, all HPF data-parallel constructs are mapped to one canonical form.

[3] Such codes are sometimes referred to as fully data-parallel codes.

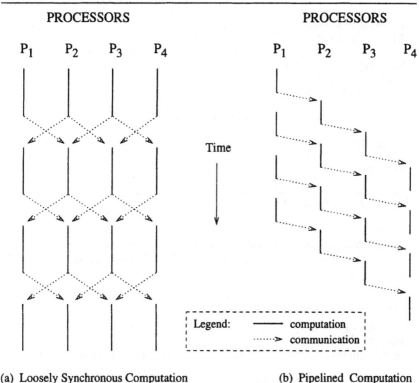

(a) Loosely Synchronous Computation   (b) Pipelined Computation

Fig. 4.4. Loosely Synchronous and Pipeline Computations

When optimizing the performance of programs, the most gains will come from optimizing the regions of the program that require the most time - iterative loops and array statements. The subsequent sections will focus on the parallelization of countable loops, where the trip count can be determined without executing the loop and adaptation of array statements to the execution on DMSs.

4.2.1 Parallelization of Loops

Loops are common in a large number of scientific codes and most Fortran programs spend most of their execution time inside loops. Therefore, loop-iteration level parallelism is one of the most common forms of parallelism being exploited by optimizing compilers for parallel computer systems.

In Fig. 4.6 below, we show a Vienna Fortran program fragment for Jacobi relaxation, together with one possible method for distributing the data: A, B and F are all distributed by **BLOCK** in both dimensions.

```
PARAMETER (M=...,N=...)
REAL, DIMENSION (1:N,1:N) :: A, B, F
      ...
DO ITIME = 1, ITIMEMAX
   DO J = 2, N-1
      DO I = 2, N-1
S:            A(I,J) = 0.25 * (F(I,J) + B(I-1,J) + B(I+1,J) + &
                             B(I,J-1) + B(I,J+1))
      ENDDO
   ENDDO
      ...
END DO
```

Fig. 4.5. Jacobi Fortran Program

In general, the set of elements of array A that are allocated in the local memory of a processor p is called the local (distribution) segment of p with respect to A.

```
PARAMETER :: M=2, N=16
PROCESSORS R(M,M)
REAL, DIMENSION(1:N,1:N), DISTRIBUTED(BLOCK,BLOCK) :: A, B, F
      ...
DO ITIME = 1, ITIMEMAX
   DO J = 2, N-1
      DO I = 2, N-1
S:            A(I,J) = 0.25 * (F(I,J) + B(I-1,J) + B(I+1,J) + &
                             B(I,J-1) + B(I,J+1))
      ENDDO
   ENDDO
      ...
END DO
```

Fig. 4.6. Jacobi Vienna Fortran Program

In Fig. 4.7, we show the SPMD form of the Jacobi code, transformed to run on a set of M^2 processors, using a pseudo notation for message passing code which has the semantics as follows.

The basic *message passing operations* in this notation are SEND and RECEIVE. Assume below that the exp_i are expressions with respective values a_i, the v_i are variables, and p_1, p_2 are processors. If

$$\text{SEND } exp_1, \ldots, exp_m \text{ TO } p_2$$

/* code for a non-boundary processor p having coordinates (P1,P2) */

PARAMETER(M = ..., N = ..., LEN = N/M)

/* declare local arrays together with overlap areas */
REAL A(1:LEN,1:LEN), F(1:LEN,1:LEN), B(0:LEN+1,0:LEN+1)

/* global to local address conversion */
/* $\widehat{f(I)}$ represents: f(I) – \$L1(P1) + 1 and $\widehat{g(J)}$ represents: g(J) – \$L2(P2) + 1 */
...
DO ITIME = 1, ITIMEMAX

/* send data to other processors */
  SEND (B(1,1:LEN)) TO PROC(P1-1,P2)
  SEND (B(LEN,1:LEN)) TO PROC(P1+1,P2)
  SEND (B(1:LEN,1)) TO PROC(P1,P2-1)
  SEND (B(1:LEN,LEN)) TO PROC(P1,P2+1)

/* receive data from other processors, assign to overlap areas in array B */
  RECEIVE B(0,1:LEN) FROM PROC(P1-1,P2)
  RECEIVE B(LEN+1,1:LEN) FROM PROC(P1+1,P2)
  RECEIVE B(1:LEN,0) FROM PROC(P1,P2-1)
  RECEIVE B(1:LEN,LEN+1) FROM PROC(P1,P2+1)

/* compute new values on local data $\lambda(p) = A(\$L1(P1):\$U1(P1),\$L2(P2):\$U2(P2))$ */
  DO I = \$L1(P1), \$U1(P1)
      DO J = \$L2(P2), \$U2(P2)
          $A(\widehat{I}, \widehat{J}) = 0.25*(F(\widehat{I}, \widehat{J})+$ &
                     $B(\widehat{I-1}, \widehat{J})+B(\widehat{I+1}, \widehat{J})+ B(\widehat{I}, \widehat{J-1})+B(\widehat{I}, \widehat{J+1}))$
      END DO
  END DO
  ...
END DO

Fig. 4.7. SPMD Form of In-Core Jacobi Relaxation Code for a Non-Boundary Processor

is executed in processor p_1, then the message $a = (a_1, \ldots, a_m)$ is sent to p_2; if

$$\text{RECEIVE } v_1, \ldots, v_m \text{ FROM } p_1$$

is executed in processor p_2, then the two statements are said to match and the transfer of the message is completed by performing the assignments $v_1 := a_1, \ldots, v_m := a_m$.

Message passing accomplishes synchronization, since a message can be received only after it has been sent. If a SEND or RECEIVE statement is delayed until a matching RECEIVE or SEND is executed, we speak of block-ing or synchronous, otherwise of non-blocking or asynchronous communication statements. In all of the following we assume an asynchronous SEND and a synchronous RECEIVE. Furthermore, we assume that a sequence of messages sent from a processor p_1 to a processor p_2 arrives in the same order in which it was sent.

We have simplified matters by only considering the code for a non-boundary processor. Further, we assume that the array sizes are multiples of M. Optimization of communication has been performed insomuch as messages have been extracted from the loops and organized into vectors for sending and receiving. Each processor has been assigned a square subblock of the original arrays. Local space of the appropriate size has been declared for each array on every processor. Array B has been declared in such a way that space is reserved not only for the local array elements which arrange the local segment (see below), but also for those which are used in local computations, but are actually owned by other processors. This extra space surrounding the local elements is known as the overlap area; this term will be explained in more details later. Values of A on the local boundaries require elements of B stored non-locally for their computation. These must be received, and values from local boundaries must be sent to the processors which need them. The work is distributed according to the data distribution: computations which define the data elements owned by a processor are performed by it - this is known as the *owner computes* paradigm. The processors then execute essentially the same code in parallel, each on the data stored locally. We preserve the global loop indices in the SPMD program. Therefore, the global to local index conversion must be provided to access the appropriate local elements of A and B. The iteration ranges are expressed in terms of the *local segment* ranges which are parameterized in terms of the coordinates of the executing processor. It is assumed that if processor p has coordinates (P1,P2) then the local segment of A on processor p, $\lambda^A(p) = A(\$L1(P1):\$U1(P1), \$L2(P2):\$U2(P2))$.

The translation of regular codes from Vienna Fortran to message passing Fortran can be described as a sequence of phases, each of which specifies a translation between two source language levels. We use the above Jacobi code to illustrate some of the transitions involved.

Assume that initially a normalized form Q_1 of a Vienna Fortran source program is given. We transform Q_1 into a target message passing Fortran program Q_4 in three conceptually distinct consecutive phases: (1) initial par-allelization, (2) optimization, and (3) target code generation.

Phase1: Initial Parallelization

Normalized Program $Q_1 \mapsto$ Defining Parallel Program Q_2

This transformation is performed by processing the data distribution specified in the original Vienna Fortran program in two steps, which are referred to as masking and communication insertion:

1. Masking enforces the owner computes paradigm by associating a boolean guard, called the mask, with each statement, in accordance with the ownership implied by the distributions: each statement S is replaced by the masked statement

    IF $(mask(S))$ THEN S ENDIF

    where
    -   $mask(S)$ is OWNED$(A(x))$ if S is an assignment statement of the form $A(x) = \ldots$, where A is partitioned and x is an associated list of subscript expressions. $OWNED(A(x))$, executed in processor p, yields true iff B is owned by p.
    -   $mask(S)\equiv$true in all other cases. Clearly, then the masked statement associated with S can be immediately replaced by S.

    Note that $mask(S)\equiv$true for all statements which are not assignments (in particular, control statements) and for all assignments to replicated variables.
2. For all non-local data accesses, communication insertion generates communication statements which copy non-local data items to private variables of the processor. For this, the *exchange primitive of level 0, EXCH0,* as defined in Fig. 4.8, is used: For every statement S with $m = mask(S)$ and every right-hand side reference *ref* to a partitioned array in S, CALL $EXCH0(m, ref, temp)$ is inserted before S. In S, *ref* is replaced by the private variable *temp*. Each execution of a SEND *ref* TO p' in a processor p, which is caused by a call to $EXCH0(m, ref, temp)$, corresponds to exactly one execution of RECEIVE *temp* FROM p in p', caused by a call to the same occurrence of $EXCH0(m, ref, temp)$, in the same state. The order of different EXCH0 statements associated with a statement S is irrelevant; however, their private variables must be distinct.

Note finally that, according to the SPMD execution model, the compiler does not generate separate node programs for each processor. Instead, each will execute the same program, receiving its parameters and initial data from special files generated at compile time.

The effect of initial parallelization of the program of Fig. 4.6 is shown in Fig. 4.9.

Phase 2: Optimization

Defining Parallel Program $Q_2 \mapsto$ Optimized Parallel Program Q_3

/* EXCH0(m,ref,temp)
The algorithm below specifies the effect of executing a call to $EXCH0(m,ref,temp)$.
Here, $MVAL(mask(S),p)$ yields true iff the value of $mask(S)$ in the current state is
true for processor p. In the case where a data item ref is owned by more than one
processor, $MASTER(ref)$ determines the processor responsible for organizing the
communication with respect to ref. */

```
IF (OWNED(ref)) THEN
    temp := ref;
    IF  (MY_PROC = MASTER(ref)) THEN
          FOR EVERY p SUCH THAT MVAL(m,p) ∧ (ref∉ λ(p))
                          SEND temp TO p
          ENDFOR
    ENDIF
ELSE
    IF (¬ OWNED(ref) ∧ MVAL(m,MY_PROC)) THEN
          RECEIVE temp FROM MASTER(ref)
    ENDIF
ENDIF
```

Fig. 4.8. The EXCH0 Primitive

The defining parallel program – as produced by Phase 1 – specifies exactly
the work distribution and communication required by the input program.
However, it would be very inefficient to actually compile and execute such a
program on a parallel computer since communication involves only single data
items and in general each processor has to evaluate the mask of a statement
for all instances of that statement.

In Phase 2, the defining parallel program is transformed into an optimized
parallel message passing Fortran program, Q_3. The efficiency of the parallel
program is improved by reducing the communication overhead and storage
requirements and increasing the amount of useful work that may be performed in parallel by mask optimization. Masks are eliminated by suitably
modifying the (processor-specific) bounds of do loops where possible. This
can be achieved in many cases by propagating the information in masks to
the loop bounds. This loop transformation is often referred to as *strip mining*.
Communication statements are moved out of loops to enable sending vector
of data and combined to perform aggregate communication where possible.
Redundant communication is eliminated. A prerequisite for many optimizations is precise flow and data dependence information, as gathered in Phase
1, and **overlap analysis**, which detects certain simple regular communication
patterns and re-organizes communication based upon this information, as explained below. Overlap analysis also helps determine the minimum amount of
storage which must be reserved for each partitioned data array in the memory of a node processor.

```
PARAMETER :: M=2, N=16
PROCESSORS R(M,M)
REAL, DIMENSION(1:N,1:N), DISTRIBUTED(BLOCK,BLOCK) :: A, B, F
PRIVATE REAL TEMP1, TEMP2, TEMP3, TEMP4
          ...
DO ITIME = 1, ITIMEMAX
    DO J = 2, N-1
        DO I = 2, N-1
            CALL EXCH0(OWNED(A(I,J)),B(I-1,J),TEMP1)
            CALL EXCH0(OWNED(A(I,J)),B(I+1,J),TEMP2)
            CALL EXCH0(OWNED(A(I,J)),B(I,J-1),TEMP3)
            CALL EXCH0(OWNED(A(I,J)),B(I,J+1),TEMP4)
            IF (OWNED(A(I,J))) THEN
S:              A(I,J) = 0.25 * (F(I,J) + TEMP1 + TEMP2 &
                                 + TEMP3 + TEMP4)
            ENDIF
        ENDDO
    ENDDO
          ...
END DO
```

Fig. 4.9. Jacobi Relaxation Code After Initial Parallelization

Overlap Analysis. Overlap analysis is performed in the compiler to determine which non-local elements of a partitioned array are used in a processor. For many regular computations, the precise pattern of non-local accesses can be computed; this information can then be used both to determine the storage requirements for the array and to optimize communication.

For each partitioned array A and processor $p \in P$, the local segment $\lambda^A(p)$ is allocated in the local memory associated with p. In the defining parallel program, communication is inserted each time a potentially non-local element of A is referenced in p – and private variables are created to hold copies of the original non-local values. Overlap analysis is used to allocate memory space for these non-local values in locations adjacent to the local segment. More precisely, the **overlap area**, $OA(A,p)$, is the smallest rectilinear contiguous area around the local segment of a process p, containing all non-local variables accessed (see Fig. 4.10). The union of the local segment and the overlap area is called the **extension segment**, $\Lambda^A(p)$, associated with A and p. This description can be used to significantly improve the organization of communication; it facilitates memory allocation and the local addressing of arrays. The relevant analysis is described in detail in [118].

The overlap area for an array A is specified by its **overlap description**, $OD(A)$, which is determined by the maximum offsets for every dimension of the local segments, over all processors. If n is the rank of A, this takes the form:

$$OD(A)=[dl_1 : du_1, \ldots, dl_n : du_n]$$

Here, dl_i and du_i denote the offsets with respect to the lower and upper bound of dimension i. Finally, the overlap description of a statement S with respect to a right-hand side reference *ref*, $OD(S,ref)$, is defined as the contribution of S and *ref* to $OD(A)$.

Example 4.2.1. Consider the example in Fig. 4.9. The overlap descriptions satisfy:

$$OD(F) = [0 : 0, 0 : 0] \text{ and } OD(B) = [1 : 1, 1 : 1]$$

while the overlap area of array F is empty, that of array B consists of an area of depth 1 around the local segment. For example, $\lambda^B(R(1,2)) = B(1 : 8, 9 : 16)$. The associated extension segment is $B(0 : 9, 8 : 17)$, and the overlap area is $B(0, 9 : 16) \cup B(9, 9 : 16) \cup B(1 : 8, 8) \cup B(1 : 8, 17)$.

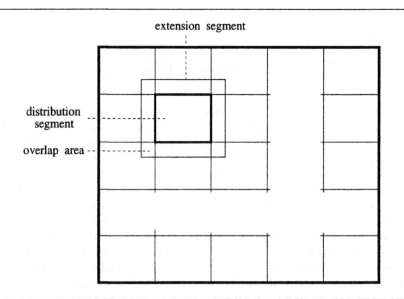

Fig. 4.10. Overlap Area for Array B in Jacobi Code ($M \geq 4$)

We outline the computation of overlap descriptions and areas by considering the one-dimensional case.

Consider an assignment statement $S : A(y) = \ldots B(x) \ldots$ in a loop with loop variable I and assume that A, B are distributed arrays and that x and y are linear expressions in I.

1. Assume that $\mu^A = \mu^B$ and $x = c_1 * I + d_1, y = c_2 * I + d_2$, where $c_1 = c_2$, and d_1, and d_2 are constants.

Let $d := d_1 - d_2$. If $d \geq 0$, then $OD(S,B(x))=[0 : d]$; otherwise $OD(S,B(x))=[-d : 0]$.

2. Assume now that $\mu^A = \mu^B$, x and y as above, but $c_1 \neq c_2$. Then, if a constant d is obtained from the symbolic subtraction $x - y$, we can proceed as above; otherwise, we apply the test below.

3. The general test is applicable to arbitrary distributions μ^A, μ^B. It constructs a set of inequalities that have to be satisfied by the loop variable for the instances of S associated with a process $p \in P$. This is based on y and the bounds of the local segments $\lambda^A(p)$. For example, if $y = c_2 * I + d_2$ and $\lambda^A(p) = B(l_p : r_p)$ then $l_p \leq c * I + d \leq r_p$ must be satisfied for each p. The corresponding set of constraints for the loop variable can be determined by applying the Fourier-Motzkin method [106, 175] to these inequalities. From these constraints, $\mu^A(p)$, and x, the overlap description $OD(S, B(x))$ can be computed.

4. Whenever there is insufficient static information to determine the value of a component in the overlap description, a conservative estimate has to be made. This may mean that the remainder of an array dimension is included in the overlap area. In the worst case, we obtain an extension segment with $\Lambda^B(p) = B$.

The Communication Primitive *EXCH*. The overlap concept can be used to organize communication: the exchange statement *EXCH0* is replaced by *EXCH*, which refers to overlap descriptions *OD(S,ref)* rather than statement masks. *EXCH* is defined in Fig. 4.11.

/* EXCH(ref,od)

Assume that S is a statement, *ref* is a right hand side reference of S of the form $A(...)$ where A is distributed, and *od=OD(S,ref)*.
MY_ PROC returns the unique identifier of the executing processor.
The algorithm below specifies the effect of executing a call to *EXCH(ref,od)*. */

```
IF ( OWNED(ref) ∧ ( MY_ PROC = MASTER(ref))) THEN
        FOR EVERY p SUCH THAT   ref ∈  OA(A,p)
                        SEND   ref TO  p
        ENDFOR
ELSE
        IF ( ref ∈  OA(A,MY_ PROC)) THEN
              RECEIVE   ref FROM   MASTER(ref)
        ENDIF
ENDIF
```

Fig. 4.11. The EXCH Primitive

The statements of the new version of *Jacobi* are given in Fig. 4.12.

```
DO  J = 2, N-1
    DO  I = 2, N-1
        CALL  EXCH(B(I-1,J),[1:0,0:0])
        CALL  EXCH(B(I+1,J),[0:1,0:0])
        CALL  EXCH(B(I,J-1),[0:0,1:0])
        CALL  EXCH(B(I,J+1),[0:0,0:1])
        IF (OWNED(A(I,J))) THEN
                S: A(I,J) = 0.25 * (F(I,J) + B(I-1, J) + B(I+1, J) + &
                                    B(I, J-1) + B(I, J+1))
        ENDIF
    ENDDO
ENDDO
```

Fig. 4.12. Loops from the Jacobi Program with EXCH-Based Communication

Optimization of Communication. The modified exchange statement forms the basis for a more advanced optimization of communication: we generalize *EXCH* to an **aggregate communication primitive** that moves blocks of data rather than single objects, and then discuss how to move communication out of loops and vectorize communication statements.

EXCH can be readily extended to move any rectilinear contiguous section $\hat{A}$ of a segment $\lambda^A(p)$ by replacing *ref* in Fig. 4.11 with $\hat{A}$, and taking care that more than one processor may be the source of a RECEIVE.

The communication for a data item can be moved out of a loop if no true dependence is violated. Loop distribution and vectorization [274] can then be applied to generate aggregate communication. Further optimization of communication includes fusion and **elimination of redundant communication**, as described in the literature [118, 120, 183]. A more general approach – independent of the overlap concept – is discussed in the next section 4.2.2.

Optimization of Masking. After initial masking, all processors execute the same masked statement sequence. For each masked statement, each processor first evaluates the mask and executes the corresponding (unmasked) statement instance if the mask yields true. The following transformations optimize the handling of masks in loops and, in many cases, lead to the strip-mining [274] of the loop across the processors by partitioning the iteration space:

1. Iteration Elimination: Iteration elimination deletes irrelevant statement instances by eliminating an entire loop iteration for a process.
2. Mask Simplification: If it can be shown that a mask is always true for each instance of each process, it can be eliminated.

A more general discussion of mask optimization and the associated work distribution can be found in [119]. We again illustrate these optimizations as well as the target code generation with the example program.

```
PARAMETER :: N=16
PARAMETER :: $LEN1(p)=$U1(p)-$L1(p)+1, $LEN2(p)=$U2(p)-$L2(p)+1
REAL  A(1:$LEN1,1:$LEN2), F(1:$LEN1,1:$LEN2)
REAL  B(0:$LEN1+1,0:$LEN2+1)
       . . .
/* global to local address conversion */
/* f̅(I̅) represents: f(I) - $L1(P1) + 1 and g̅(J̅) represents: g(J) - 1 */
       . . .
DO  ITIME = 1, ITIMEMAX
    CALL  EXCH(B(1:N-2,2:N-1),[1:0,0:0])
    CALL  EXCH(B(3:N,2:N-1),[0:1,0:0])
    CALL  EXCH(B(2:N-1,1:N-2),[0:0,1:0])
    CALL  EXCH(B(2:N-1,3:N),[0:0,0:1])
    DO  J = MAX(2,$L1(p)),MIN(N-1,$U1(p))
        DO  I = MAX(2,$L2(p)),MIN(N-1,$U2(p))
            A(I̅, J̅) = 0.25*(F(I̅, J̅)+ &
                      B(I̅-1, J̅)+B(I̅+1, J̅)+ B(I̅, J̅-1)+B(I̅, J̅+1))
        ENDDO
    ENDDO
       . . .
END DO
```

Fig. 4.13. Jacobi – Final Version: Code for Processor p

Fig. 4.13 gives the final version of the Jacobi iteration. The program is parameterized in terms of the executing processor, p. It is assumed that $\lambda^A(p) = B(\$L1(p) : \$U1(p), \$L2(p) : \$U2(p))$. Hence, for example, $\$L2(R(1,2)) = 9$ and $\$U2(R(1,2)) = 16$. The local declarations reserve space for the extension segment of B. The execution of the first exchange statement, CALL $EXCH(B(1:N-2,2:N-1),[1:0,0:0])$, in a processor p has the following effect: first, all elements of $B(1:N-2,2:N-1)$ that are owned by p and belong to the corresponding overlap area of another processor p', are sent to p'. Secondly, all elements of $B(1:N-2,2:N-1)$ that are in $OA(B,p)$ are received from the respective owner. For example, $R(2,1)$ sends $B(9:14,8)$ to $R(2,2)$, and $R(1,2)$ receives $B(1:8,8)$ from $R(1,1)$. These communication operations can be executed in parallel.

For each p, S is executed exactly for those values of I and J that satisfy $MAX(\$L1(p),2) \le I \le MIN(\$U1(p), N-1)$ and $MAX(\$L2(p),2) \le J \le MIN(\$U2(p))$. For these iterations, the mask can be eliminated; all other iterations can be eliminated for p.

Phase3: Target Code Generation

Optimized Parallel Program $Q_3 \mapsto$ Message Passing Target Program Q_4

In the final phase the internal representation of the optimized parallel program is transformed into the Fortran 90 message passing program that is adapted to the target architecture.

4.2.2 Adaptation of Array Statements

Array statements are core constructs of Vienna Fortran 90 and HPF. They enable the programmer to express data parallelism in a natural way. When processing an array statement the task of the compiler is to match the parallelism expressed in the statement to that of the target massively parallel architecture. This subsection briefly describes techniques developed for adapting and optimizing array statements for distributed-memory machines. The description of these techniques is introduced in more detail in [23, 25, 22].

In the following paragraphs, we describe the compilation model applied for processing array assignment statements and introduce basic and optimizing transformations.

Basic Model and Terminology

The parallelization process as described in this subsection is based on the $SPMD$ model of computation. Starting from the user specified data alignment and data distribution directives the compiler has to determine the layout of the data on the set virtual of processors and to determine how to spread the work among the processors available. Work distribution is achieved by applying the *owner computes* rule, so that each processor only computes those data elements that are allocated in its local memory. Access to non-local data is handled via explicit message passing.

In the following we assume that all declarations of objects (i.e. arrays and processor arrays) are normalized such that the lower bound in each dimension is 1. Fig. 4.14 summarizes the notation used in this subsection.

A, B	names of arrays	$\mathbf{I}^A$	index domain of array A
N	size of array dimension	i	global index, $1 \leq i \leq N$
M	parameter for **BLOCK**(M) or	$f(i)$	function, mapping indices of
	CYCLIC(M) distributions		*lhs* array to corresponding
P	total number of processors		indices of *rhs* array
	in dimension	$[l : u : s]$	regular array section
p	individual processor, $1 \leq p \leq P$	Δ^A	distribution descriptor

Fig. 4.14. Table of symbols used in the Description of Parallelization

Data Layout. The set of indices of an array A which are mapped to a particular processor p by means of a distribution function μ is called the *local index set* of p with respect to A and denoted by $Local^A(p)$.

$$Local^A(p) = \{i \mid \mu(i) = p \wedge i \in \mathbf{I}^A\}$$

The distribution functions corresponding to the BLOCK (M) and CYCLIC (M) distributions are summarized in Fig. 4.15.

BLOCK (M)	CYCLIC (M)
$\mu(i) = \left\{\left\lceil \frac{i}{M} \right\rceil\right\}, M \geq \left\lceil \frac{N}{P} \right\rceil$	$\mu(i) = \left\{1 + \left(\left\lceil \frac{i}{M} \right\rceil - 1\right) \bmod P\right\}$

Fig. 4.15. Distribution Functions

A distribution for a multi-dimensional array is specified by describing the distribution of each array dimension separately, without any interaction of dimensions.

The local index set of an array is the basis for the computation of execution sets and communication sets and is also needed for the transformation of array declarations.

Due to the nature of the BLOCK (M) and CYCLIC (M) distributions, the local index set of an array dimension can be represented as a section by means of triplet notation. For BLOCK, CYCLIC (1) and BLOCK (M) distributed dimensions, the local index set can be described by a single triplet. This, however, might be not the case for CYCLIC (M) distributions, where the local index set may consist of more than one triplet, which complicates the handling of these sets at compile- and/or at runtime.

The local index set of an array A is specified (in global indices) by formulas in Fig. 4.16. In the case of a CYCLIC (M) distribution with $M < \left\lceil \frac{N}{P} \right\rceil$, some processors will get more than one section of array elements. The total number of sections on a particular processor p is given by n_p.

Index Conversion. Given an array $A(1 : N)$ with a BLOCK (M) or CYCLIC (M) distribution onto P processors, $1 \leq p \leq P$, and a global index i of A(i), then the corresponding local index l on processor p is determined as shown in Fig. 4.17.

Work distribution. As already mentioned, work distribution is derived on basis of the *owner computes* paradigm. The set of elements of the *lhs* variable[4] which have to be computed on a particular processor p is referred to as *execution set*, denoted by $exec(p)$. If the *lhs* variable of the assignment statement

---

[4] left hand side variable

BLOCK (M) Distribution: **CYCLIC (M) Distribution:**

$Local^A(p) =$
$[(p-1)*M+1 : min(p*M, N) : 1]$
$$Local^A(p) = \bigcup_{i=1}^{n_p} \sigma_i(p) = [\mathcal{L}_i(p) : \mathcal{U}_i(p) : 1]$$

$$\sigma_i(p) = [\mathcal{L}_i(p) : \mathcal{U}_i(p) : 1]$$

$$n_p = \left\lceil \frac{N - (p-1)*M}{P * M} \right\rceil$$

CYCLIC Distribution:

$$\mathcal{L}_i(p) = (p-1)*M+1+(i-1)*P*M$$

$Local^A(p) = [p : N : P]$

$$\mathcal{U}_i(p) = min(\mathcal{L}_i(p) + M - 1, N)$$

Fig. 4.16. Local Elements of an Array A

BLOCK (M) Distribution   **CYCLIC (M) Distribution**

$g2l(i) = 1 + (i-1) \bmod M$ $g2l(i) = (i-1) \bmod M + M * \left\lfloor \frac{i-1}{M*P} \right\rfloor + 1$

$p = \left\lceil \frac{i}{M} \right\rceil, \ 1 \leq i \leq N, \ M \geq \left\lceil \frac{N}{P} \right\rceil$ $p = 1 + \left(\left\lceil \frac{i}{M} \right\rceil - 1 \right) \bmod P, \ 1 \leq i \leq N$

Fig. 4.17. Index Conversion

---

is a regular array section, the execution set can be represented using Fortran 90 triplet notation.

Let for an array assignment statement of the form $A(l : u : s) = ...$, $\mathbf{L} = [l : u : s]$ denote the so-called *lhs reference space* of array A. The execution set according to a an assignment to a regular array section $A(l : u : s)$ on a particular processor p can be determined by the intersection of the *lhs* reference space $\mathbf{L}$ with the local index set on this processor.

$$exec_{\mathbf{L}}^A(p) = Local^A(p) \cap [l : u : s]$$

For **BLOCK (M)** distributions the execution set can always be described via a single section whereas for **CYCLIC (M)** distributions this might not be the case. The summary of closed form expressions for $exec(p)$ can be found in [23, 22].

Communication Sets. When executing the SPMD form of an assignment statement that references distributed arrays, each processor has to determine the set of non-local elements it must receive from other processors in order to perform all the computations defined by its execution set. Furthermore, it also has to determine which elements of its local element set have to be sent to other processors. **Communication sets** determine those elements a particular processor p must send (or receive) to (from) another processor q, in order to compute the elements corresponding to its execution set set.

---

for each processor q, $q \neq p$

1. generate the execution set for processor q, $exec_L^A(q)$
2. determine $f(exec_L^A(q))$, where $f(i) = \frac{(i - l_{l_1})}{s_{l_1}} * s_{r_1} + l_{r_1}$
3. compute the intersection $f(exec_L^A(q)) \cap Local^B(p)$ which corresponds to $send\_set^B(p, q)$

where $\mathbf{L} = [l_{l_1} : u_{l_1} : s_{l_1}]$.

Fig. 4.18. Communication Sets for Array Assignments

Consider an assignment statement of the form $A(\text{lhs}) = B(\text{rhs})$, where, lhs $= [l_{l_1} : u_{l_1} : s_{l_1}]$ and rhs $= [l_{r_1} : u_{r_1} : s_{r_1}]$. The set of elements a processor p must send to a processor q is called *send_set(p,q)*. It comprises those elements of B that are in p's local memory and that are needed by another processor q. The set of non-local elements a processor p needs from a processor q is called *recv_set(p,q)*. Note that *recv_set(p,q)* is given by *send_set(q,p)*. In the case of array assignment statements that only reference regular sections, computation of communication sets is based on regular section intersection.

Computation of communication sets for an array assignment statement of the form shown above is summarized in Fig. 4.18.

Processing Array Statements

There are two main steps in the initial parallelization of an array statement: (1) *Masking and Execution Set Splitting.* Array assignment statements are masked (guarded) to ensure that all variable updates are exclusively in the local memory of the executing processor. The execution set of the statement is split into the local part (local iteration set) that uses only data that is local on the executing processor, and the non-local part (non-local iteration set) which uses some non-local data stored in the communication buffer (see the transformation of the array statement in Fig. 4.20). The splitting also enables the forthcoming optimizing transformation to achieve overlapping communication and computation.

(2) *Communication Insertion.* Possible data movements between processors are specified by means of so-called *communication descriptors* COMM *s* which are generated and inserted before this statement, for every *rhs* reference that may cause communication.

High Level Communication Descriptor COMM. A communication descriptor contains all information that is needed for the calculation of execution sets and communication sets. Communication descriptors are flexible enough to allow for: recognizing and removing redundant communication descriptors, movement of descriptors (i.e. extraction from loops, movement across procedure boundaries), fusion of communication descriptors, splitting into the

Send Part

if $Local^B(p)$ ∩ rhs ≠ φ
      ∧ p = MASTER($Local^B(p)$) then
   for each q ∈ P, q ≠ p do
      if $send\_set^B(p,q)$ ≠ φ) then
                  M = B($send\_set^B(p,q)$)
                  send M to q
      endif
   endfor
endif

Receive Part

if $Local^A(p)$ ∩ lhs ≠ φ then
   for each q ∈ P, q ≠ p,
            p = MASTER($Local^B(q)$) do
      if ($recv\_set^B(p,q)$ ≠ φ) then
         receive M from q
      endif
   endfor
endif

Fig. 4.19. Specification of the COMM Descriptor

sending and receiving components. Moreover, they enable a uniform treatment of references to scalars, array elements, and array sections. So optimization of communication is performed entirely on the level of communication descriptors. In the final phase which adapts the SPMD program for the target machine, communication descriptors are transformed into the explicit message passing form.

Let **lhs**, **rhs** denote section subscripts and let Δ^A, Δ^B denote distribution descriptors of arrays A and B, respectively(each distributed array declared is represented by a descriptor). Let MM denote a temporary buffer, and let P denote the set of available processors, with arbitrary processors $p, q \in P$. An array communication descriptor of the form

$$\boxed{\text{COMM}\,(\Delta^A, < \text{lhs} >, \Delta^B, < \text{rhs} >, M)}$$

which for example is generated in case of an array assignment statement

$$A(\textbf{lhs}) = B(\textbf{rhs})$$

has the semantics introduced in Fig. 4.19 when evaluated on a particular processor p. In this figure, $MASTER(s)$ is a function which returns a uniquely defined processor m which ownes the array element set s. This processor is responsible for sending the data. The choice of m is system dependent. In many cases the overlap area communication approach can be used. Then M will denote the overlap area description and the **COMM** will have the same semantics as the $EXCH$ communication primitive specified in Subsection 4.2.1.

Original code

```
REAL , A(N), B(M)
REAL , DISTRIBUTED  μ^A :: A(N)
REAL , DISTRIBUTED  μ^B :: B(M)

A(lhs) = B(rhs)
```

Transformed code

```
COMM (Δ^A, < lhs >, Δ^B, < rhs >, tmp_B)
! local accesses to B
WHERE  (Owned_Arr(A(lhs)) .AND. Owned_Arr(B(rhs)))
        A(lhs) = B(rhs)
! non-local accesses to B
WHERE  (Owned_Arr(A(lhs)) .AND..NOT. Owned_Arr(B(rhs)))
        A(lhs) = tmp_B
```

Fig. 4.20. Transformation of the Array Assignment Statement

Transformation of Array Assignments. Basic transformations applied to an array assignment statement of type $A(\mathbf{lhs}) = B(\mathbf{rhs})$ are shown in Fig. 4.20.

The inquiry function $Owned\_Arr(ref)$ tests the locality of the array elements referenced. It uses an array or array section as an input parameter and produces a result which is a logical array with the same shape as ref. The test is evaluated for each array element referenced by ref, and the corresponding position in the function result array is set to true, if that element is *owned* by the processor calling this function.

If more than one distributed array appears on the right hand side, a communication descriptor has to be generated for each of those arrays. However, in the case of multiple array references appearing on the right hand side, splitting of the array assignment statement into local and non-local parts is more complex.

Masked Array Assignment. The goal of basic transformations applied to **WHERE** statements[5] is to avoid communication for arrays involved in the mask evaluation at the time when the assignment statement is executed.

---

[5] We assume that **WHERE** constructs are transformed into a series of **WHERE** statements during the program normalization phase.

Therefore it is necessary to enforce that the mask is aligned with the array on the left hand side of the array assignment statement. This is achieved by generating a temporary mask array, which is aligned with the array on the left hand side. Prior to the WHERE statement, an assignment to the temporary mask array is inserted.

Optimizations

In this phase, the compiler applies transformations to improve performance of the resulting SPMD program. The program analysis must capture sufficient information to enable sophisticated optimizations. Data flow analysis information, for example, enables to decide whether execution and communication sets computed for one array statement can be reused for other ones.

Transformed code

```
COMMS (Δ^A, < lhs >, Δ^B, < rhs >, tmp_B)
! computations accessing only local elements of B
! - this hides the latency
WHERE (Owned_Arr(A(lhs)) .AND. Owned_Arr(B(rhs)))
        A(lhs) = B(rhs)

COMMR (Δ^A, < lhs >, Δ^B, < rhs >, tmp_B)
! computations accessing non-local elements of B
WHERE (Owned_Arr(A(lhs)) .AND..NOT. Owned_Arr(B(rhs)))
        A(lhs) = tmp_B
```

Fig. 4.21. Hiding Latency by Local Computation

There are several compiler optimizations which can reduce or hide the latency overhead. Communication latency consists of two additive components: $T_c = T_s + T_t$, where, $T_s [\mu sec]$ is the communication startup time, and $T_t [\mu sec]$ is the message transmission time. The effect of T_s can be reduced by minimizing the total number of messages sent. This can be achieved by the following techniques:

- *Elimination of redundant communication statements.* The communication statement is redundant if all data elements communicated by this statement are communicated in the same way by another communication statement.
- *Communication statement fusion.* Messages generated from different communication statements performing data transmission for the same array can be combined.
- *Collective communication.* Message overhead can be reduced by utilizing fast collective communication routines instead of generating individual

messages, when communication takes place between groups of processors in regular patterns.

Latency can also be reduced using an approach that is often called the *latency tolerance* [154]. The aim is to *hide* T_t, the message transmission time, by overlapping communication with parallel computation. This strategy permits to hide the latency of remote accesses. The execution set for each processor is split into local and non-local parts. Each processor executes separately parts of expressions which access only local data and those which access non-local data, as expressed in Fig. 4.21. COMM statements are split into send (COMMS) and receive (COMMR) parts whose semantics are clear from the specification introduced in Fig. 4.19. COMMS is moved before the part performing local computations, and COMMR between this part and the one performing computations in which accesses to copies of non-local data are needed.

BIBLIOGRAPHICAL NOTES

In this part, a number of systems is discussed which have been constructed to compile code for distributed memory systems.

Most of the work described here is based upon Fortran or C, although there are a few systems based upon functional languages.

SUPERB [118, 271, 272] is a semi-automatic parallelizer for standard Fortran 77 programs, initially developed at the University Bonn. SUPERB was the first implemented restructuring tool of this kind and supports both automatic parallelization as well as vectorization. The system takes, in addition to a sequential Fortran program, a specification of a desired data distribution and converts the code to an equivalent program to run on a distributed memory machine (Intel iPSC, the GENESIS machine, and SUPRENUM), inserting the communication required and optimizing it where possible. SUPERB provides special support for handling work arrays, as are commonly used in Fortran codes, for example to store several grids in one array. Besides data dependence analysis and interprocedural data flow analysis the system performs interprocedural distribution and overlap analysis, interprocedural distribution propagation, procedure cloning.

The Vienna Fortran Compilation System (VFCS) [27], which is a further development of SUPERB currently handles both Vienna Fortran and subset HPF programs. Furthermore, all the language constructs designed for the specification of parallel I/O operations are being currently implemented. The VFCS can be used as a command line compiler for Vienna Fortran or HPF programs, or as an interactive parallelization system where the parallelization process is guided by user decisions. The VFCS currently generates message passing code for the iPSC/860, the Intel Paragon, the Meiko CS-2, and PARMACS [36] and MPI [112], programming interface for parallel computers.

The Kali compiler [165, 167] was the first system that supported both regular and irregular distributions. From the data distribution specification the compiler generates functions to determine the set of elements owned by each processor. The distribution of forall loop iterations is represented by means of execution sets. Based on the set of local elements and the execution set, communication sets are determined. In the absence of enough compile time information, these sets have to be computed at runtime. By examining simple array access patterns, Koelbel [166] has worked out a lot of special cases which can be handled efficiently.

The Fortran D compiler [146, 248] converts subset Fortran D programs into SPMD node programs with explicit message passing calls. The compiler is integrated with the ParaScope programming environment which supplies information from data flow and dependence analysis and provides support for loop transformations. Several communication optimization techniques like communication vectorization, vector message pipelining, message coalescing, message aggregation and exploitation of collective communication have been investigated [145, 140].

The ADAPT system [188] transforms *Distributed Fortran 90* (DF90) into standard Fortran 77 programs with embedded calls to the ADLIB library [68], which provides communication routines and customized Fortran 90 intrinsic procedures. The parallelization process is based on the SPMD programming model and aims to exploit the parallelism inherent in array assignments.

The Forge 90 system [152] is an interactive parallelization system for Fortran targeting shared nad distributed memory MIMD machines. It provides tools that perform data dependence analysis, data flow analysis and interprocedural analysis. The input language is subset-HPF with additional restrictions.

The Adaptor system [49] is a prototype system for interactive source-to-source transformation for Fortran programs. The input language is Fortran 77 with Fortran 90 array extensions, and data distribution directives. The system has no facilities for automatic parallelization (i.e. no dependence analysis) and only exploits the parallelism inherent in array assignment statements.

The source language for the Crystal compiler [75, 182, 183] is the functional language Crystal, which includes constructs for specifying data parallelism. Experimental compilers have been constructed for the iPSC hypercube and the nCUBE. The program is transformed in a number of phases, first to a shared memory form and thence to a distributed memory form.

Dino [224, 225] represents an early attempt to provide higher-level language constructs for the specification of numerical algorithms on distributed memory machines. It is explicitly parallel, providing a set of C language extensions. The DINO compiler generates C code with iPSC/2 message-passing constructs.

Dataparallel C [141] is a SIMD extension of the C language. The compilers of this language have been constructed for both shared and distributed memory machines; they produce C code with calls to the Dataparallel C routing library.

The Pandore system [13] takes as input a sequential C program, annotated with a virtual machine declaration and data distribution specifications, and generates parallel C code augmented with calls to message passing primitives.

Other currently existing systems include commercial HPF compilers: PGI compiler [211], NAS Mapper [198], PREPARE compilation system [252], etc.

4.3 Parallelization of Irregular Codes

The automatic parallelization of *irregular* codes is a challenging problem of great practical relevance. Irregular codes can be found in the number of industrial and research applications. From the computer science point of view, an application is considered to be irregular if there is the limited compile time knowledge about data access patterns and/or data distributions. In such an application, access patterns to major data arrays are only known at runtime. Furthermore, major data structures are accessed through one or more levels of indirection, via some form of index arrays, which requires runtime preprocessing and analysis in order to determine the data access patterns. On the other hand, in *regular* problems, data distributions and access patterns may be described using an expression easily recognizable by a compiler (e.g., stencils [179]).

Application areas in which irregular codes are found include unstructured computation fluid dynamics, molecular dynamics, time dependent flame modeling, diagonal or polynomial preconditioned iterative linear solvers, etc.

In irregular codes, the array accesses cannot be analyzed at compile time to determine either independence of these or to find what data must be prefetched and where it is located. Therefore, the appropriate language support, as well as the compile time techniques relying on runtime mechanisms, are needed.

Vienna Fortran and HPF-1, provide several constructs to express data parallelism in irregular codes explicitly[6]. These include the FORALL statement and construct, the DO loop prefixed by the INDEPENDENT directive, and the array statements. Further extensions for *computation control* provided recently by HPF-2 allow the explicit *work distribution* specification in forall loops via an *ON clause*, and the control of reductions in such loops. When processing these constructs, the task of the compiler is to match the parallelism expressed by the construct to that of the target parallel system.

In this section, we discuss the computational domain and structure of irregular applications and introduce the programming environment provided by the VFCS for irregular applications. In detail, we describe the basic techno-

---

[6] The Vienna Fortran support for irregular computations was described in Sections 3.2.4 and 3.2.6.

logy that we have developed for parallelization of irregular applications. A great deal of these techniques has already been implemented either within the VFCS or PREPARE compiler [53]. Generality and computational and memory utilization efficiency have been major design objectives in this research and development. Experimental performance results taken from the real application kernel are also included in this section.

4.3.1 Irregular Code Example: Computation on an Unstructured Mesh

As described in the previous paragraph, irregular problems are characterized by indirect data accesses where access patterns are only known at runtime. The input data represent a description of a discretized computational domain which is irregular.

The technique of applying a system of equations over a discretized domain which is used, for example, in computational mechanics[7], leads inevitably to the concept of a *mesh* or *grid*. The complexity of a computational grid ranges from the simple regular structured to fully unstructured.

Irregular and block structured grids, suitable for transport phenomena modeling, are widely used in the development of finite volume (finite difference/control volume) schemes for computational fluid dynamics. Unstructured grids were introduced to allow finite volume methods to work with arbitrarily complex geometries.

Unstructured grid codes are unlikely to offer the computational efficiency of structured grid codes. The implicit nature of a structured grid avoids the need of indirection in variable addressing and allows great efficiency of coding, cache utilization and vectorization. On the other hand, unstructured grids provide a far greater flexibility for the modeling of complex geometries and avoid the need for the complexity of a block structured code.

An unstructured grid is specified as a hierarchy of components or grid entities, each of which may be regarded as a data object or structure which can be used to provide a spatial (geometric) or topological (connectivity) reference to the variables used in the code. The definition of an unstructured grid begins with a set of grid points or nodes, each of which is defined by a set of spatial coordinates. The grid points describe the geometric shape and physical size of the examined object (computational domain). From a collection of nodal points a grid element can be constructed. Nodal points are connected by a set of edges or faces which represent the linkage information among nodes or elements.

Fig. 4.22 illustrates a typical irregular loop. The loop *L2* in this code represents a sweep over the edges of an unstructured mesh of size: NNODE, NEDGE, where NNODE is the number of data elements (nodes) and NEDGE

---

[7] There it is used for modeling of diverse physical systems (structural mechanics, structural dynamics, electromagnetics, magnetohydrodynamics, etc.).

```
INTEGER, PARAMETER :: NNODE = ..., NEDGE = ...
REAL x(NNODE), y(NNODE)                      ! data arrays
INTEGER edge1(NEDGE), edge2(NEDGE)           ! indirection arrays
      . . .
L1:   DO j = 1, nstep                        ! outer loop
L2:      DO i = 1, NEDGE                      ! inner loop
            x(edge1(i)) = x(edge1(i)) + y(edge2(i))
            x(edge2(i)) = x(edge2(i)) + y(edge1(i))
         END DO
      END DO
```

Fig. 4.22. An Example with an Irregular Loop

represents the number of edges describing connections among the nodes. Fig. 4.23 depicts a structure of a simple unstructured mesh that will be used for illustration of our examples.

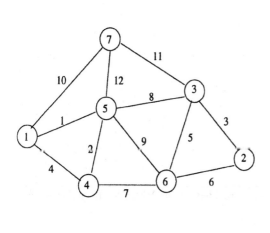

i	edge1(i)	edge2(i)
1	1	5
2	5	4
3	2	3
4	1	4
5	5	2
6	2	6
7	6	4
8	5	3
9	6	5
10	1	7
11	7	3
12	5	7

Fig. 4.23. Example Mesh with Seven Nodes and Twelve Edges

Since the mesh is unstructured, indirection arrays have to be used to access the vertices during a loop sweep over the edges. The reference pattern is specified by the integer arrays *edge1* and *edge2*, where *edge1(i)* and *edge2(i)* are the node numbers at the two ends of the *i*th edge. Arrays *edge1* and *edge2* are called *indirection arrays*. The calculation on each node of the mesh requires data from its neighboring nodes. The arrays *x* and *y* represent the values at each of the *NNODE* nodes; these arrays are called *data arrays*.

Such a computation forms the core of many applications in fluid dynamics, molecular dynamics, etc.

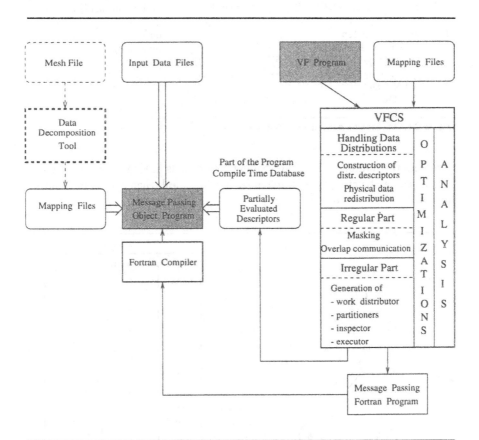

Fig. 4.24. Programming Environment Provided by the VFCS for Irregular Applications

4.3.2 Programming Environment for Processing Irregular Codes

This part outlines the structure and functionality of the environment provided by the VFCS for irregular applications. We focus on a class of irregular problems that can be expressed by explicit parallel (forall) loops.

Several parts of the VFCS apply to both regular and irregular features of the application, and some parts are specific to either kind. Fig. 4.24 gives an overview how the subsystem implementing irregular codes is embedded into VFCS and integrated with the whole processing environment. The regular part of VFCS was discussed in Section 4.2. The functionality of other components of this environment is explained below.

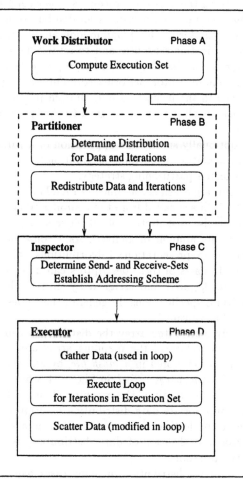

Fig. 4.25. Computing Phases Generated for a FORALL Loop

The process of transforming irregular array references combines compile-time parallelization techniques with runtime analysis. Processing strategy based on the *inspector/executor* paradigm [165, 228] is used to implement runtime analysis and parallel execution. Each forall loop is transformed into three main code segments as depicted in Fig. 4.25: the *work distributor*, *inspector*, and *executor*. The work distributor determines how to spread the work (iterations) among the available processors – on each processor, it computes the *execution set*, i.e. the set of iterations to be executed on this processor. The inspector performs the dynamic loop analysis of the forall loop to determine communication schedules and to establish an appropriate addressing scheme for the access to local and to copies of non–local elements on each processor. The executor performs communication according to he schedules determined in the inspector, and, on each processor, executes the actual computations for all iterations assigned to that processor in the work distribution phase.

The user can optionally specify the application of a *partitioner* for each forall loop, which calculates the optimal distribution for data and iterations, according to the selected partitioning strategy.

The essence of this approach is contained in PARTI [95] and CHAOS [217] runtime routines[8] which are called from the compiler embedded code.

Generally, the precise distribution of all objects cannot be determined at compile time by any amount of analysis. As a consequence, it is necessary to have a runtime representation of distributions, called *data distribution descriptors*. Only parts of these descriptors can be determined at compile time – they are declared and assigned by initial values according to ALIGN/DISTRIBUTE specifications. In general, the values of the decriptor items are calculated at runtime.

For an irregularly distributed array the distribution can be specified by a mapping array. The values of the mapping array are usually computed from the mesh description stored in a *mesh file*, by an external partitioner, referred to as *Data Decomposition Tool* in Fig. 4.24, which stores them in a file called *mapping file*. At runtime, the mapping file is used as input for the construction of an appropriate data distribution descriptor. If the mapping file is available at compile time, it can be treated as a special source file and used for partial initialization of data distribution descriptors and for optimizations of the program.

A new dynamic array distribution in many cases requires not only the updating the values of the data distribution descriptor for that array, but the *physical array redistribution* as well.

Compile time *optimizations* are supported by the *analysis* component of the VFCS. Extensive compile time analysis enables production of highly efficient code. Some important optimization techniques are introduced in Section 4.5.

---

[8] CHAOS is a runtime library that subsumes PARTI.

4.3.3 Working Example

A part of the parallelization techniques will be illustrated on the example given in Fig. 4.26. It is derived from the unstructured mesh linear equation solver GCCG which is a part of the program package FIRE. Loop L2 represents the kernel loop of this solver.

```
        PARAMETER :: NP = ..., NNINTC =..., NNCELL = ...
        PROCESSORS P1D(NP)
        DOUBLE PRECISION, DIMENSION (NNINTC),                    &
            DISTRIBUTED ( BLOCK ) :: D2, BP, BS, BW, BL, BN, BE, BH
        INTEGER, DISTRIBUTED ( BLOCK, :) :: LC(NNINTC,6)
S1:     DOUBLE PRECISION, DYNAMIC :: D1(NNCELL)
S2:     INTEGER, DISTRIBUTED ( BLOCK ) :: MAP(NNCELL)
        ...
S3:     READ (u) MAP
S4:     DISTRIBUTE D1 :: ( INDIRECT (MAP))
        ...
        step = 1
L1:     DO WHILE (condition .AND. (step .LE. MAXSTEPS))
        ...
L2:     FORALL nc = 1, NNINTC ON OWNER (D2(nc))
            D2(nc) = BP(nc) * D1(nc) −                           &
                     BS(nc) * D1(LC(nc, 1)) −                    &
                     BW(nc) * D1(LC(nc, 4)) −                    &
                     BL(nc) * D1(LC(nc, 5)) −                    &
                     BN(nc) * D1(LC(nc, 3)) −                    &
                     BE(nc) * D1(LC(nc, 2)) −                    &
                     BH(nc) * D1(LC(nc, 6))
        END FORALL
        ...
        step = step+1
        END DO
```

Fig. 4.26. Kernel Loop of the GCCG Solver

FIRE[9] is a general purpose computational fluid dynamics program package [15] developed specially for computing compressible and incompressible turbulent fluid flow as encountered in engineering environments. For the discrete approximation the computational domain is subdivided into a finite number of control volumes – *internal* cells. For boundary conditions, an additional layer of infinitely thin cells is introduced around the computational domain, called *external* cells. Flow variables and coefficients associated with each cell are held in one–dimensional arrays of two different sizes: one size corresponding to the number of internal cells (parameter NNINTC), and the second size corresponds to the total number of cells, internal plus external

---

[9] FIRE is a registered trade-mark of AVL LIST Gmbh.

cells (parameter NNCELL). These kinds of arrays are related with each other in such a way that the first portion of the bigger arrays (including internal cells), is aligned with the smaller arrays. To determine the interconnection of cells, an indirect addressing scheme is used. Description of the interconnection of cells is stored in the two–dimensional array LC, where the first index stands for the actual cell number and the second index denotes the direction to its neighboring cell. All calculations are carried out over the internal cells, external cells can only be accessed from the internal cells, no linkage exists between the external cells.

Array $D1$ is indirectly distributed (statement $S4$) using the mapping array MAP which is initialized from a dynamic mapping file (statement $S3$); the remaining arrays have got the **BLOCK** distribution. The mapping file is constructed separately from the program through the *Domain Decomposition Tool (DDT)* [109]. The DDT inputs the computational grid corresponding to the set of internal cells (see Fig. 4.24), decomposes it into a specified number of partitions, and allocate each partition to a processor. The resulting mapping array is then extended for external cells in such a way that the referenced external cells are allocated to that processor on which they are used, and the non–referenced external cells are spread in a round robin fashion across the processors. In Section 4.3.10 we compare performance of the GCCG code which includes the above code fragment with performance of a version that only includes regular distributions. This version is simply derived from Fig. 4.26 by replacing the keyword **DYNAMIC** in statement $S1$ by the distribution specification: **DISTRIBUTED(BLOCK)**, and removing statements $S2$, $S3$ and $S4$. The main loop $L2$ within the outer iteration cycle computes a new value for every cell by using the old value of this cell and of its six indirectly addressed neighbored cells[10]. This loop has no dependences and can be directly transformed to the parallel loop. Work distribution **ON OWNER(D2(nc))** ensures that communications are caused only by the array $D1$ because other arrays (BP, BS, BW, BL, BN, BE, and BH) have the same distribution and access patterns as array $D2$.

4.3.4 Distribution Descriptors

For each irregularly distributed array A and in each processor p, the *constructor of data distribution descriptors (CDDD)* creates the runtime data distribution descriptor $DD^A(p)$ which includes information about: shape, alignment and distribution, associated processor array, structure and size of the local portion of A in processor p. In particular, it includes the local index set $Local^A(p)$ of the array A and the translation table $Trat^A$. These terms are defined below (we repeat some definitions from Subsection 4.2.2 to provide a more compact presentation of the compilation model).

---

[10] The remaining loops, with exception of the one that updates the $D1$, perform either calculations on local data only or global operations (global sums).

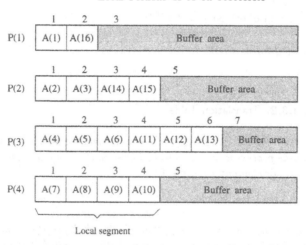

Data Distribution Specification

PROCESSORS P(4)
INTEGER, PARAMETER :: MAP(16) = (/1,2,2,3,3,3,4,4,4,4,3,3,3,2,2,1)/
REAL, DISTRIBUTED (INDIRECT(MAP)) :: A(16)

Fig. 4.27. Local Portions

Definition 4.3.1. Local Index Set and Local Segment

The set of indices of an array A which are mapped to a particular processor p by means of a distribution function is called the local index set *of p with respect to A and denoted by $Local^A(p)$. An array element $A(i)$, $i \in I^A$, is said to be owned by processor p, iff $i \in Local^A(p)$; p is the* home processor *of $A(i)$. The set of elements of A that are allocated in the local memory of a processor p is called the* local segment *of p with respect to A and is denoted by $LSeg^A(p)$.*

The local index set of an array is the basis for the generation of the appropriate translation table.

INTEGER :: MAP(local^{MAP}_size)
INTEGER, DIMENSION (:), POINTER :: local^{D1}
INTEGER :: local^{D1}_size
TYPE (TT_EL_TYPE), DIMENSION (:), POINTER :: tt^{D1}
TYPE (DIST_DESC) :: dd^{D1}

      . . .
C−−Constructing local index set for D1 using the MAP values
      CALL build_Local(dd^{MAP}, MAP, local^{D1}, local^{D1}_size)
C−−Constructing translation table for D1
      tt^{D1} = build_TRAT(local^{D1}, local^{D1}_size)
C−−Constructing runtime distribution descriptor for D1
      dd^{D1}%local = local^{D1}
      dd^{D1}%local_size = local^{D1}_size
      dd^{D1}%tt = tt^{D1}
      ... initialization of other fields of dd^{D1} ...

Fig. 4.28. CDDD for Array D1

Definition 4.3.2. Translation Table

Translation table for an array A, Trat^{A}, is a distributed data structure which records the home processor and the local address in the home processor's memory for each array element A(i), i ∈ $\mathbf{I}^{A}$.

Definition 4.3.3. Local Portion

The local memory of a processor p that is allocated for the local segment of A and the corresponding communication buffer is called the local portion of A with respect to p. The local portion is of the following structure:

$$(\, LSeg^{A}(p) \, ; \; \mathcal{B}^{A}(p))$$

where $\mathcal{B}^{A}(p)$ denotes the communication buffer. The buffer space follows immediately the local segment data.

Allocation of local portions and their structuring into local segments and buffers is illustrated in Fig. 4.27.

Only one distribution descriptor is constructed for a set of arrays having the same distribution. Usually a distribution descriptor is initialized whenever an array is associated with a new distribution. For every INDIRECT distribution the distribution descriptor can be constructed as soon as the values of the corresponding mapping array are defined. The construction of a distribution descriptor must precede the program point at which the DISTRIBUTE statement is specified.

In Fig. 4.28, the construction of the distribution descriptor for the array *D1* from the working example in Fig. 4.26 is depicted. On each processor p, the procedure build_Local computes the local index set and its cardinality, denoted by variables $local^{D1}$ and $local^{D1}\_size$, respectively. These two objects are input arguments to the PARTI function build_TRAT that constructs the translation table for *D1* and returns the pointer, tt^{D1}, to it.

4.3.5 Physical Data Redistribution

If the NOTRANSFER attribute is not given for the array A within a *distribute* statement, the physical transfer of the array elements is needed after the CDDD phase to achieve the data layout corresponding to the new distribution. This physical remapping is achieved by a call to the runtime procedure Remap which is built up on top of PARTI:

CALL Remap $(dd_{old}^A, dd_{new}^A, A_{loc})$

where dd_{old}^A and dd_{new}^A are the old and new distribution descriptors of A, respectively, and A_{loc} is the address of the local portion of A on the calling processor.

4.3.6 Work Distributor, Inspector and Executor

Work Distributor

On each processor p, the work distributor computes the *execution set* $exec(p)$, i.e. the set of loop iterations to be executed on processor p. Depending on the work distribution specification, the compiler selects different ways for determination of $exec(p)$. Their overview is presented in Fig. 4.29.

In the case of *on owner* and *on processor* work distribution, the inversion functions (g and f in Fig. 4.29), in general, may not exist. Then an algorithmic method has to be used. Efficient parallel algorithms for the computation of execution sets are introduced in [229].

For the loop *L2* in the example shown in Fig. 4.26, the execution set is determined in a simple way:

$$exec(p) = [1 : NNINTC] \cap local^{D2}(p) = local^{D2}(p)$$

Execution Sets for Tightly Nested Forall Loops

Methods presented in Fig. 4.29 can also be generalized for computation of execution sets of tightly nested forall loops. Consider, for example, a normalized n–nested forall loop with the header:

FORALL $(i_1 = l_1, u_1, \ldots, i_n = l_n, u_n)$ ON OWNER $(A(f_1(i_1), \ldots, f_n(i_n)))$

The parameters in the formulas have the following meaning:

NP ... number of processors
p ... number of the individual processor, $0 \le p \le (NP-1)$
N ... size of the global iteration set, $N = U - L + 1$
$[L:U]$ ... a contiguous range of integers $\{ i \mid L \le i \le U \}$

FORALL i = L,U
  DISTRIBUTED (BLOCK)
Iterations are divided into contiguous blocks, whose sizes differ by at most 1.
Let $q = \lfloor \frac{N}{NP} \rfloor, r = MOD(N, NP)$.
If $N < NP$ then
$exec(p) = \{p + L\}$, for $p < N$
$exec(p) = \phi$, for $p \ge N$
else
$exec(p) = [q * p + L : q * (p+1) + L]$,
         for $p < r$
$exec(p) = [q * p + L : q * (p+1) - 1 + L]$,
         for $p \ge r$

FORALL i = L,U
  ON OWNER (A(f(i)))
Iteration i is executed on the processor which owns the array element A(f(i)).
If f is an invertible function then
$$exec(p) = f^{-1}((local^A(p)) \cap [L:U])$$
where, $local^A(p)$ denotes the set of local indices of the array A on processor p.

FORALL i = L,U
  ON PD1(g(i))
Iteration i is executed on the processor $g(i)$. If g is an invertible function then
$$exec(p) = g^{-1} \cap [L:U]$$

FORALL i = L,U
  DISTRIBUTED (CYCLIC)
Iterations are distributed in a round robin fashion across the processors.
$$exec(p) = [p + L : U : NP]$$

Fig. 4.29. Strategies for Determination of the Execution Set

where $A(L_1 : U_1, \ldots, L_n : U_n)$ denotes an n–dimensional array that is distributed onto an m–dimensional processor array R. Furthermore, let particular processor $p \in R$ be designated by the m–tuple $p = (p_1, \ldots, p_m)$, and i-th dimension of A by the A_i, for $i = 1, \ldots, n$.

Then the set of local indices $local^A(p)$ of a multi–dimensional array A is given by:

$$local^A(p) = local^{A_1} \times \ldots \times local^{A_n}$$

where for $i = 1, \ldots, n$

$$local^{A_i} = \begin{cases} local^{A_i}(p_j) & \text{if i--th dimension of A is distributed} \\ & \text{onto processor dimension j} \\ [L_i : U_i] & \text{otherwise} \end{cases}$$

The execution set $exec^A(p)$ is given by:

$$exec^A(p) = dim\_exec^{A_1} \times \ldots \times dim\_exec^{A_n}$$

where for $i = 1, \ldots, n$

$$dim\_exec^{A_i} = \begin{cases} exec^{A_i}(p_j) = f_i^{-1}((local^{A_i}(p_j)) \cap [l_i : u_i]) \\ \quad \text{if } i\text{-th dimension of } A \text{ is distributed} \\ \quad \text{onto processor dimension } j; \\ \quad (\text{if } f_i^{-1} \text{ does not exist, use the} \\ \quad \text{appropriate algorithmic method}) \\ \\ [l_i : u_i] \qquad \text{otherwise} \end{cases}$$

The sets $dim\_exec^{A_i}$, for $i = 1, \ldots, n$, are called *dimensional execution sets*. The declaration of the derived type DIM_ESD for a dimensional execution set can be of the form as introduced in Fig. 4.30. If the dimensional execution set may be expressed by a regular section, it is represented by a triple. On the other hand, irregular array sections are represented by one–dimensional integer arrays. This approach speeds–up accesses to regular execution sets.

```
TYPE DIM_ESD           ! dimensional execution set descriptor
   LOGICAL isreg       ! indicates whether dimensional execution set is regular
   INTEGER l,u,s       ! regular section description
   INTEGER size        ! size
   INTEGER, DIMENSION(:), POINTER :: dim_es   ! dimensional execution set
END TYPE DIM_ESD
...
TYPE (DIM_ESD) :: esd(loop_nest_depth)
```

Fig. 4.30. Data Structures Designed for Storing Execution Sets

Inspector and Executor

The semantics of the Vienna Fortran forall loop guarantees that the data needed during the loop is available before its execution begins. Similarly, if a processor modifies the data stored on another one, the updating can be deferred until execution of the forall loop finishes.

The inspector performs a dynamic loop analysis. Its task is to describe the necessary communication by a set of so called *schedules* that control runtime procedures moving data among processors in the subsequent executor phase. The dynamic loop analysis also establishes an appropriate addressing scheme to access local elements and copies of non–local elements on each processor. Information needed to generate a schedule, to allocate a communication buffer, and for the global to local index conversion of array references can be produced from the appropriate global reference lists, along with the knowledge of the array distributions.

Global reference list is constructed by collecting access patterns through a do loop which iterates over the computed execution set. Optimizing symbolic analysis makes sure that only the textually distinct access patterns are included into the list.

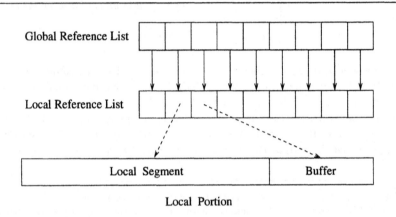

Fig. 4.31. Index Conversion by Localize Mechanism

In the case of a multi–level indirection, that means, a distributed array is referenced through other distributed arrays, communication may be required to form the global reference list. So, a *multi–phase inspector/executor* starting with the innermost subscripts must be applied. Processing the array access patterns is carried out in multi–phases, where the order guarantees that non-local data referenced are resolved in the previous phases.

Local reference list is used to reference distributed array elements in the loop execution phase. The addressing scheme utilized supposes the buffer for non–local elements to follow immediately the local segment of the array. Local reference list is constructed in such a way that the local index points either to the local segment or to the buffer. This is called the *localize mechanism* and its principles are graphically sketched in Fig. 4.31. The kernel for the implementation of the localize mechanism and generation of communication schedules is the **Localize** procedure whose interface is specified in Fig. 4.32.

Communication schedule contains the necessary information to identify the locations in distributed memory from which data is to be acquired. Global reference list is scanned using a hash table in order to remove any duplicate references to non–local elements, so that only a single copy of each non–local datum is transmitted.

```
subroutine Localize (dd, ndata, glob_ref, loc_ref, nnonloc, sched)
    type(dist_desc), intent(in) :: dd              ! distribution descriptor
    integer, intent(in) :: ndata                   ! number of references
    integer, dimension(:), intent(in) :: glob_ref  ! list of the global references
    integer, dimension(:), intent(out) :: loc_ref  ! list of inearized local references
    integer, intent(out) :: nnonloc                ! number of nonlocal references
    type (schedule), pointer :: sched              ! schedule (ouput argument)
end subroutine Localize
```

Fig. 4.32. Interface for Localize

The **executor** is the final phase in the implementation of the forall loop; it performs communication described by schedules, and executes the actual computations for all iterations in $exec(p)$. The schedules control communication in such a way that execution of the loop using the local reference list accesses the correct data in local segments and buffers. Non–local data that are needed for, or updated in the computations on processor p are gathered from, or scattered to other processors via the runtime communication procedures which accept a schedule, an address of the local segment, and an address of the buffer[11] as input arguments. The generic interface of the runtime communication procedure **Gather** is given by Fig. 4.33. Interface for Scatter has the same form. The forall loop body is transformed so that every subscript of a distributed array is replaced by the corresponding local reference.

```
interface Gather
    subroutine fGather(buf, locseg, sched)        ! specific version for real arrays
        real, dimension(:) :: buf                  ! buffer
        real, dimension(:) :: locseg               ! local segment
        type(schedule), pointer :: sched           ! schedule
    end subroutine fGather

    ... interface bodies for other specific versions ...
end interface Gather
```

Fig. 4.33. Generic Interface for Gather

The inspector and executor generated for the loop in the example given in Fig. 4.26 are introduced in Fig. 4.34. Global reference lists $globref_i$, $i = 1, \ldots,$

---

[11] If necessary, space for the local segment and buffer is reallocated prior to the Gather call, depending on the current size of the segment and the number of non–local elements computed by Localize.

C **INSPECTOR phase**

C Constructing the execution set exec by the work distributor
$$\ldots$$
es_size= exec(1)%size;   es_low= exec(1)%l;   es_high= exec(1)%u

C Constructing global reference list for the 1st dimension of LC
```
G1:     n1 = 1
G2:     DO k = es_low, es_high
G3:        globref1(n1) = k;   n1 = n1 + 1
G4:     END DO
```

C Index Conversion for LC having the data distribution descriptor dd1
```
I1:     CALL Ind_Conv (dd1, n1-1, globref1, locref1)
```

C Constructing global reference lists for D1, D2, BP, BS, BW, BL, BN, BE, BH
```
G5:     n1 = 1;   n2 = 1;   n3 = 1
G6:     DO k = es_low, es_high
G7:        globref2(n2) = k;   n2 = n2 + 1
G8:        globref3(n3) = k
G9:        globref3(n3+1) = LC(locref1(n1),1)
G10:       globref3(n3+2) = LC(locref1(n1),4)
G11:       globref3(n3+3) = LC(locref1(n1),5)
G12:       globref3(n3+4) = LC(locref1(n1),3)
G13:       globref3(n3+5) = LC(locref1(n1),2)
G14:       globref3(n3+6) = LC(locref1(n1),6)
G15:       n3 = n3 + 7;   n1 = n1 + 1
G16:    END DO
```

C Index Conversion for D2, BP, BS, BW, BL, BN, BE, BH
C   having the common data distribution descriptor dd2
```
I2:     CALL Ind_Conv (dd2, n2-1, globref2, locref2)
```

C Computing schedule and local reference list for D1
```
I3:     CALL Localize (dd3, n3-1, globref3, locref3, nonloc3, sched3)
```

C **EXECUTOR phase**

C Gather non-local elements of D1
```
        nbuf = dd3%local_size+1
E1:     CALL Gather (D(nbuf), D(1), sched3)
```

C Transformed loop
```
        n2 = 1;   n3 = 1
        DO k=1, es_size
          D2(locref2(n2)) = BP(locref2(n2)) * D1(locref3(n3)) −         &
                            BS(locref2(n2)) * D1(locref3(n3+1)) −       &
                            BW(locref2(n2)) * D1(locref3(n3+2)) −       &
                            BL(locref2(n2)) * D1(locref3(n3+3)) −       &
                            BN(locref2(n2)) * D1(locref3(n3+4)) −       &
                            BE(locref2(n2)) * D1(locref3(n3+5)) −       &
                            BH(locref2(n2)) * D1(locref3(n3+6))
          n2 = n2 + 1;   n3 = n3 + 7
        END DO
```

Fig. 4.34. Inspector and Executor for the GCCG Kernel Loop

3, are computed from the subscript functions and from $exec(p)$[12] (see lines
G1–G16). The *globref$_i$*, its size and the distribution descriptor of the referenced array are passed to the PARTI procedure Localize (line *I3*) to determine
the appropriate schedule *sched$_i$*, the size *nonloc$_i$* of the communication buffer
to be allocated, and the local reference list *locref$_i$* which contains results of
global to local index conversion. Declarations of arrays in the message passing code allocate memory for the local segments holding the local data and
the communication buffers storing copies of non–local data. The procedure
Ind_Conv (lines *I1*, *I2*) performs the global to local index conversion for array references that refer only local elements. In the executor phase, non–local
elements only of the array *D1* must be gathered by the runtime PARTI procedure Gather (line *E1*); the work and data distribution specifications guarantee
that all other array elements are accessed locally.

4.3.7 Handling Arrays with Multi–Dimensional Distributions and General Accesses

The PARTI library is able to handle arrays with only one–dimensional
BLOCK and CYCLIC(1) distributions. On the other hand, a PARTI library
extension called *PARTI+* and the associated compiling techniques support
all Vienna Fortran and HPF distributions and alignments.

Organization of communication and addressing scheme for arrays with
multi–dimensional distributions is based on viewing the array as a one–
dimensional array whose distribution is determined by the dimensional distributions of the original array. Let's consider the code introduced in Fig.
4.35.

PROCESSORS $PnD(M_1, M_2, \ldots, M_n)$, $P1D(M_1 * M_2 * \ldots * M_n)$
REAL, DISTRIBUTED $(dist_1, \ldots, dist_n)$ TO PPnD :: $A(N_1, \ldots, N_n)$
REAL, DISTRIBUTED (LDIST $(n, dist_1, \ldots, dist_n))$ &
                  TO PP1D :: $LA(N_1 * \ldots * N_n)$
EQUIVALENCE (A, LA)

Fig. 4.35. Multi–Dimensional/One–Dimensional Distribution Transformation

If the references to *A* are to be handled by runtime techniques, the compiler introduces the declaration of a new one–dimensional distributed array
LA and transforms all references to *A* into references to *LA*. The distribution
of *LA* is calculated from the dimensional distributions $dist_1, \ldots, dist_n$ by the
distribution function[13] LDIST. The communication buffer is appended to the

---

[12] In Fig. 4.34, $exec(p)$ and its cardinality are denoted by the variables *exec* and
exec_size respectively.
[13] Remember that in VF90, user-defined distribution functions may be specified.

local segment of the linearized array. Communication precomputation and communication are supported by special runtime procedures. This approach is described in more detail in [53].

```
PARAMETER :: N1= ... , N2= ... , M1= ... , M2= ...
PROCESSORS  P(M1,M2)
REAL , DIMENSION (N1,N2), DISTRIBUTED ( CYCLIC, BLOCK) TO P :: A
REAL , DIMENSION (N1,N2), DISTRIBUTED ( BLOCK, CYCLIC) TO P :: B
    ...
      FORALL  (i=1:N1, j=1:N2)
S:        A(fl(i), gl(j)) = B(fr₁(i), gr₁(j)) * B(fr₂(i), gr₂(j))
      END FORALL
```

The subscripts are formed from functions that may contain array references.

Fig. 4.36. Vienna Fortran Source Code

Example:

Let's take the Vienna Fortran code fragment depicted in Fig. 4.36 as a running example. Possible forms of the assignment S in Fig. 4.36 that are treated by our techniques include:

1. A(fl(i), gl(j)) = C(fr(i))
2. A(fl(i), gl(j)) = D(fr(i), 4, gr(j))
3. A(fl(i), gl(j)) = B(v(i,j), w(i,j))

where A, B, C and D are arrays with one- or multi-dimensional distributions.

Work Distributor

In Subsection 4.3.6, we specified the functionality of the work distributor for non-nested forall loops. Remember that on each processor p, the work distributor computes the *execution set* $exec(p)$, i.e. the set of loop iterations to be executed in processor p. If for our example in Fig. 4.36 the work distributor applies the *owner–computes rule* as the work distribution strategy, $exec(p)$ is given by: $exec(p) = \{(i, j) \mid (fl(i), gl(j)) \in local^A(p)\}$, where $local^A(p)$ denotes the set of indices of elements of array A that are stored on p.

In Fig. 4.37, the work distributor code implementing the basic algorithm for the computation of $exec(p)$ is introduced.

PARTI+ and Inspector/Executor Code Generation

The inspector generated for our example is introduced in lines $I1$–$I13$ of Fig. 4.39. In the first step, the list $glob(p)$ of global references to B is computed *(lines I6–I12)*. This list and the distribution descriptor of B are used by the PARTI+ procedure Cview to determine the communication pattern, i.e. a *schedule*, the number of non–local elements, and the local reference list *loc(p)* which contains results of global to local index conversion. The interface for Cview is the same as the interface of Localize introduced in Fig. 4.32, and a graphical sketch of how the routine Cview works is depicted in Fig. 4.38.

```
PARAMETER :: ldepth=2                         ! loop nest depth
TYPE (DIST_DESC) :: dd^A                       ! distribution descriptor of A
INTEGER :: exec_size                           ! cardinality of the execution set
INTEGER, DIMENSION(:,:), ALLOCATABLE :: exec
...
ALLOCATE (exec(dd^A%local_size, ldepth))
exec_size = 0
DO i=1, N1
    DO j=1, N2
        IF (Is_Local(dd^A, fl(i), gl(j))) THEN
            exec_size = exec_size + 1
            exec(exec_size, 1) = i
            exec(exec_size, 2) = j
        END IF
    END DO
END DO
```

Fig. 4.37. Basic Work Distributor

$i_1 = exec(1,1)$, $j_1 = exec(1,2)$, ..., $l_{ess} = exec(ess,1)$, $J_{ess} = exec(ess,2)$
(where $ess = exec\_size$)

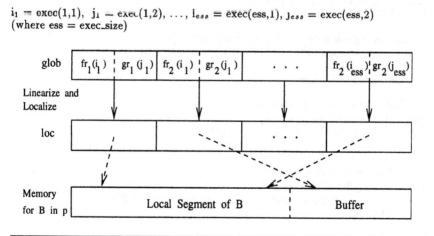

Fig. 4.38. Index Conversion by Localize Mechanism

```
TYPE (DIST_DESC) :: dd^A, dd^B          ! distribution descriptors
INTEGER :: ndim^B                        ! rank of B
INTEGER :: nref^B                        ! number of references to B
INTEGER :: exec_size                     ! size of exec
INTEGER, DIMENSION (:,:), ALLOCATABLE :: exec, ref
INTEGER, DIMENSION (:), ALLOCATABLE :: glob, loc
...
```

C   **INSPECTOR** phase

C   Constructing the execution set exec
```
        l = 0
        DO (i=1:N1, j=1:N2)
          l = l + 1
          ref(l,1) = fl(i);   ref(l,2) = gl(j)
        END DO
I1:     CALL build_ES_owner (2, 1, N1, 1, N2, dd^A, ref, exec, exec_size)
```

C   Constructing global reference list for B
```
I2:     ndim^B = 2
I3:     nref^B = 2
I4:     ndata = exec_size * nref^B
I5:     ALLOCATE (glob(ndim^B * ndata), loc(ndata))

I6:     l = ndim^B * nref^B
I7:     DO k = 0, exec_size-1
I8:        glob(k*l + 1) = fr_1 (exec(k+1, 1))
I9:        glob(k*l + 2) = gr_1 (exec(k+1, 2))
I10:       glob(k*l + 3) = fr_2 (exec(k+1, 1))
I11:       glob(k*l + 4) = gr_2 (exec(k+1, 2))
I12:    END DO
```

C   Computing schedule and local reference list for B
```
I13:    CALL Cview (dd^B, ndata, glob, loc, nnonloc, sched)
```

C   **EXECUTOR** phase

C   Gather non-local elements of B
```
E1:     nbuf = dd^B%local_size+1
E2:     CALL Gather (B(nbuf), B(1), sched)
```

C   Transformed loop
```
E3:     DO l = 0, exec_size-1
E4:        A( g2l(dd^A, fl(exec(l+1, 1) ), gl(exec(l+1, 2)))) =     &
                  B(loc(l * nref^B + 1)) * B(loc(l * nref^B + 2))
E4:     END DO
```

Fig. 4.39. Inspector/Executor Code for Arrays with Multi–Dimensional Distributions

The declaration of array B in the message passing code allocates memory for the local segment holding the local data of B and the communication buffer storing copies of non–local data. The buffer is appended to the local segment. An element from $loc(p)$ refers either to the local segment of B or to the buffer. A local index in $loc(p)$ is derived from the corresponding elements of $glob(p)$.

The executor is introduced in lines $E1$–$E4$ of Fig. 4.39. The g2l(dd^A, x, y) function call determines the offset of the element A(x,y) in the local segment of A associated with the calling processor.

4.3.8 Runtime Support

Much of the complexity of the irregular distributions implementation is contained in the VFCS_lib runtime library which is an enhanced version of the PARTI ([95]) and CHAOS ([217]) runtime libraries.

PARTI routines were originally developed as a runtime support for a class of irregular problems which consist of a sequence of concurrent computational phases, where patterns of data access and computational cost of each phase cannot be predicted until runtime. They were designed to ease the implementation of irregular problems on distributed memory parallel architectures by relieving the user of having to deal with many low–level machine specific issues. PARTI routines can be used in tasks such as global–to–local index mappings, communication schedule generation, and schedule based communication. This functionality can be used directly to generate inspector–executor code. CHAOS runtime library is a superset of the PARTI designed to partitioning data automatically. It provides efficient support for parallelization of adaptive irregular problems where indirection arrays are modified during the course of the computation.

VFCS_lib includes also procedures determined for computing local index sets, execution set, and a number of procedures providing a suitable interface between the target Fortran program and the runtime routines.

4.3.9 Optimizations

Optimization of irregular codes includes compile time and runtime features. The following paragraphs briefly discuss a list of optimizations which should be addressed by HPF compiler developers.

Compiler Analysis.
Compile time program analysis provides a basis for the determination of

– where execution set descriptors and communication schedules are to be generated and deleted, and

– where gather, scatter, and accumulate operations are to be placed.

These issues are discussed in more detail in Section 4.5.

Light–Weight Schedules.
In certain highly adaptive problems, such as those using particle–in–cell methods, data elements are frequently moved from one set to another during the course of the computation. The implication of such adaptivity is that preprocessing for a loop must be repeated whenever the data access pattern of that loop changes. In such applications a significant optimization in schedule generation can be achieved by recognizing that the semantics of set operations imply that elements can be stored in sets in any order. This information can be used to build much cheaper *light–weight* communication schedules [150]. During schedule–generation, processors do not have to exchange the addresses of all the elements they will be accessing with other processors, they only need to exchange information about the number of elements they will be appending to each set. This greatly reduces the communication costs in schedule generation.

Renumbering Mesh Nodes.
If a compiler does not support irregular distributions, it can indirectly support such distributions by reordering array elements to reduce communication requirements. This technique is described in [216]. First, a partitioner maps irregularly array elements to processors. Next, data array elements are reordered so that elements mapped to a given processor are assigned to consecutive locations. The indirection arrays are then renumbered. The reordered data arrays are then distributed blockwise.

Schedule Time–Stamping.
For codes with a deep nesting of procedure calls in a parallel loop, compile time redundancy analysis may not produce adequate results even if using an interprocedural strategy. In order to eliminate this drawback, a runtime technique called *schedule time–stamping* [215] can be used. When using this technique, the compiler generates code that at runtime maintains a record of (potential) modifications of index arrays used to indirectly reference another distributed array. In this scheme, each inspector checks this runtime record to see whether any indirection arrays may have been updated since the last time the inspector was invoked.

Program Slicing.
Some application codes use extremely complex access functions. For example, a chain of distributed array indexing may be set up where values stored in one distributed array are used to determine the indexing pattern of another

distributed array, which in turn determines the indexing pattern of a third distributed array.

In such a situation, the conventional single inspector–executor scheme breaks down because the inspector needs to make a potentially non–local reference. One possible solution is a multi–phase inspector–executor scheme: an inspector requiring communication is split into an inspector and an executor. Applying this technique recursively, we end up with multiple inspection phases, each peeling off one level of non–local access. Each phase creates opportunities for aggregated data prefetching for the next one.

Fig. 4.40 shows three way of constructing a reference to a distributed array x. All these references require two inspection phases, assuming arrays ia, ib, and ic are aligned with the loop iterations. The parallelization method that was described in this section is only able to handle the reference patterns depicted by example (a) where the level of indirection is forced by the use of subscripted values in subscripts.

```
DO i = 1, n            DO i = 1, n            DO i = 1, n
  ... x(ia(ib(i))) ...   IF (ic(i)) THEN        DO j=ia(i),ia(i+1)
END DO                     ... x(ia(i)) ...        ... x(ia(j)) ...
                         END IF                  END DO
                       END DO                  END DO

      (a)                    (b)                    (c)
```

Fig. 4.40. References Requiring Two Inspector Phases

A general transformation method for handling loops with multiple levels of indirection that was proposed by Saltz et al. [96, 97] breaks up each loop whose references require multi-phase inspectors into multiple loops whose references require single inspectors only. Each non-local reference is then a distributed array indexed by a local index array. A central concept of this approach is *program slicing* [261] on the subscript expressions of the indirect array accesses.

Exploiting Spatial Regularity.
Many real applications include irregular access patterns with certain structure. If the compiler and runtime support is able to detect and utilize this structure, computational and communication performance and memory utilization efficiency can be improved significantly [17]. In many cases this structure is reflected in the array subscript expressions as shown in the example below:

```
FORALL I=1,N
  DO J=1,R(I)
      ...
      ...B(Y(I)+J)...
  END DO
END FORALL
```

where B is a distributed data array and Y is an indirection array. Here, references to array B will have first an irregular access $B(Y(I))$ followed by a sequence of $R(I)$ contiguous accesses.

4.3.10 Performance Results

This section presents the performance of the automatically parallelized AVL FIRE benchmark solver GCCG (orthomin with diagonal scaling). The solver and its kernel loop were described in Subsection 4.3.3 and the kernel used as working example in the subsequent subsections. The generated code was optimized in such a way that the inspector was moved out of the outer iteration cycle, since the communication patterns do not change between the solver iterations.

Number of internal cells: 13845				
Number of total cells: 19061				
Number of solver iterations: 1				
Parallel code				
Number of Processors	Time (in secs)			
	Data Distr: BLOCK		Data Distr: INDIRECT	
	Inspector	Executor	Inspector	Executor
4	0.63	14.84	0.85	6.62
8	0.41	11.54	0.60	5.04
16	0.25	9.11	0.35	3.97

Table 4.1. Performance Results of the GCCG Kernel Loop

We examined the GCCG using the input dataset Tjunc, with 13845 internal cells and 19061 total number of cells. Two versions of the codes have been elaborated to evaluate the influence of different kinds of data distributions on parallel–program execution time.

In the first version, all data arrays are distributed by BLOCK, and in the second version the INDIRECT distribution was specified for the data arrays accessed in the irregular code parts. The mapping array used for the indirect distribution was determined by an external partitioner. The program has been executed on the Intel iPSC/860 system, the sequential code on 1 processor, the parallel code on 4, 8, and 16 processors. Table 4.1 depicts the timings obtained for the kernel loop in one iteration step. For the whole GCCG benchmark, where the calculation stopped after the 338 iterations, the timing results are summarized in the Table 4.2.

Number of internal cells: 13845		
Number of total cells: 19061		
Number of solver iterations: 338		
Sequential code		
Number of Processors	Time (in secs)	
1	31.50	
Parallel code		
Number of Processors	Time (in secs)	
	Data Distr: BLOCK	Data Distr: INDIRECT
4	30.25	27.81
8	24.66	19.50
16	22.69	13.50

Table 4.2. Performance Results of the GCCG Solver

BIBLIOGRAPHICAL NOTES

Solutions of many special cases of the problems discussed in this section appeared in the literature. Koelbel [165] introduces the term inspector/executor for processing irregular data–parallel loops. The developments in the runtime support and compiler support effort can be characterized as follows.

Runtime Support for Irregular Applications

The CHAOS/PARTI library [217], and in particular, the original PARTI library [95], had a significant impact in the design of other runtime supports and compilation techniques for irregular applications.

Multiblock PARTI [5] provides support for block–structured applications with regular distributions. Unknown problem sizes may disable compile time analysis for regular distributions and this library can optimize communication at runtime for those cases.

LPARX [169] is a C++ library that provides runtime support for dynamic, block–structured, irregular problems in a variety of platforms.

Compilation of Irregular Applications

KALI [167] was the first compiler system to provide support for regular and irregular applications in multi–computers. Support for irregular computations was given by the inspector/executor model, although the data structures used to keep non–local references were different from PARTI.

On top of PARTI, Joel Saltz and coworkers developed a compiler for a language called ARF (ARguably Fortran). The compiler generates automatically calls to the PARTI routines from distribution annotations and *distributed loops* with an *on clause* for work distribution [269]. ARF only supports parallelization of programs written in a small subset of Fortran. It also follows the inspector/executor model.

SUPERB [57] was the first system to automatically parallelize real irregular Fortran 77 applications. The Fortran DO loops, parallelization objects, were interactively selected (marked) by the user. PARTI was used as the runtime support. This implementation was the basis for the more advanced one within the VFCS [63, 64, 62].

Hanxleden [256, 257] developed the irregular support for the Fortran D compiler. The irregular runtime support that was used was CHAOS/PARTI. He also proposed using value–based distributions that use the contents of an array to decide on data distribution, as opposed to index–based distributions commonly used for irregular problems.

Bozkus et al [47] developed inspectors for Fortran90D forall statements containing indirection arrays. They also used PARTI as the runtime support.

At the Technical University of Zürich a compiler called OXYGEN has been developed for the K2 distributed memory multiprocessor [226]. The input language includes distribution annotations for data and loop iterations. The entire analysis of communication among nodes is done at runtime. No compile time techniques are supported.

Brezany et al. [56] proposed and implemented automatic parallelization of irregular applications for virtual shared memory systems implemented on top of the Intel iPSC/2 and compared the performance with the implementation based on the distributed–memory model.

4.4 Coupling Parallel Data and Work Partitioners to the Compiler

In Chapter 3, we presented the language constructs provided by Vienna Fortran (VF) to deal efficiently with irregular codes. These include constructs for

```
INTEGER, PARAMETER :: NN = ..., NE = ...
REAL, DYNAMIC :: x(NN), y(NN)                          ! data arrays
INTEGER, DISTRIBUTED ( BLOCK ) :: ia(NE), ib(NE)  ! indirection arrays
INTEGER, DISTRIBUTED ( BLOCK ) :: map(NN)         ! mapping array

    . . .
DISTRIBUTE ( INDIRECT (map)) :: x, y
    . . .
L1:    DO j = 1, nstep                                 ! outer loop
L2:      FORALL i = 1, NE                               ! inner loop
           x(ia(i)) = x(ia(i)) + y(ib(i))
         END FORALL
       END DO
```

Fig. 4.41. An Example with an Irregular Loop

specifying irregular distributions and asynchronous parallel loops (FORALL loops).

A typical irregular loop considered there had a form that is illustrated in Fig. 4.41. The data access pattern is determined by indirection arrays *ia* and *ib*. At runtime, once the data access patterns are known, a two phase *inspector/executor* strategy can be used to parallelize such a loop–nest. The loop in Fig. 4.41 is sometimes called a *static* irregular loop [150], since the data access pattern of the inner loop (*L2*) remains unchanged through all iterations of the outer loop (*L1*). Our basic parallelization technology for this type of irregular codes was presented in Section 4.3.

The language features discussed so far enable the user to assume full control of data and work distributions. However, in this section we show that it is also possible to provide a higher level language support and an interactive interface which enable the user to direct the system to derive data and work distributions automatically at runtime, according to a selected strategy. Then, before transforming the loop into the inspector–executor, a partitioning phase is generated which at runtime redistributes data arrays and iterations in an irregular way to minimize communication and balance load.

This approach has to be used, for example, when the number of processors that the parallel program will be running on, or the problem size are unknown at the program development time. In that case, it is not possible to use any external partitioner[14]. Furthermore, runtime partitioning is extraordinarily important for some irregular applications that are *adaptive*, in the sense that the data access patterns may change during computation. In adaptive irregular problems, such as adaptive fluid dynamics and molecular dynamics codes, interactions between entities (mesh points, molecules, etc.) change during computation (due to the mesh refinement or movement of

---

[14] A partitioner decomposes a dataset to a number of partitions (subdomains) and allocates each partition (subdomain) to each processor.

molecules). Since interactions are specified by indirection arrays, the adaptivity of irregular programs is represented by the frequency of modification on indirection arrays. In highly adaptive problems, data arrays need to be redistributed frequently.

```
L1:   DO n = 1, nstep                          ! outer loop
L2:       FORALL i = 1, ind_arrays_size        ! inner loop
              x(ia(i)) = x(ia(i)) + x(ia(i)) * y(ib(i))
          END FORALL
S:        IF (required) THEN                    ! under certain conditions
              Regenerate(ic(i)))                ! indirection array may change
          END IF
L3:       FORALL i = 1, ic_size                 ! inner loop
              x(ic(i)) = x(ic(i)) + y(ic(i))
          END FORALL
      END DO
```

Fig. 4.42. A Code that Adapts Occasionally

Fig. 4.42 illustrates properties of loops found in molecular dynamics codes and unstructured fluid dynamics codes [150]. Here, multiple loops access the same data arrays but with different access patterns. In loop $L2$ the data arrays x and y are indirectly accessed using arrays ia and ib. In loop $L3$ the same data arrays are indirectly accessed using indirection array ic. The data access pattern in loop $L2$ remains static, whereas the data access pattern in loop $L3$ changes whenever the indirection array ic is modified. The adaptivity of the loop is controlled by the conditional statement S.

In Section 4.3, we specified four phases which implement an irregular loop at runtime. These phases were outlined in Fig. 4.25 (page 104), and they are recalled in Fig. 4.43. The first two phases, A and B, concern mapping data and computations onto processors. Phase C concerns analyzing data access patterns in loops and generating optimized communication calls. Phase D carries out communication derived in phase C and performs the computation prescribed by the loop. In static irregular programs, phase D is typically executed many times, while phases A through C are executed only once. In some adaptive programs where data access patterns change periodically but reasonable load balance is maintained, phases C and D must be repeated whenever the data access patterns change. In highly adaptive programs, the data arrays may need to be repartitioned in order to maintain load balance. In such applications, all the phases B, C and D are repeated.

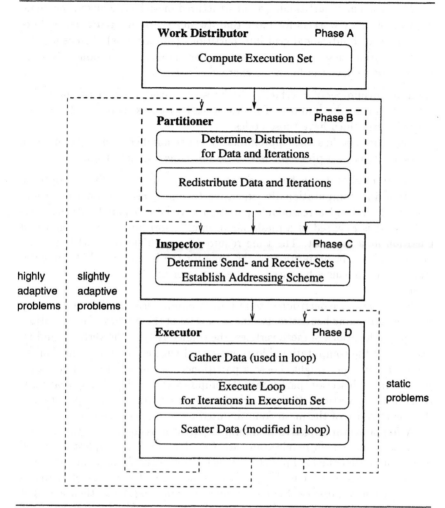

Fig. 4.43. Computing Phases Generated for a Forall Loop

4.4.1 Towards Parallel Data and Work Partitioning

Many irregular applications are based on computations on unstructured meshes. Therefore, the data and work partitioning issues we are going to discuss in this subsection will be related to the mesh concepts.

A simple unstructured mesh was depicted in Fig. 4.23 on page 102. Physical values (such as velocity and pressure) are associated with each mesh vertex. These values are called *flow variables* and are stored in data arrays. Calculations are carried out using loops over the list of edges that define the connectivity of the vertices. The distribution of data arrays is determined by the distribution of mesh points.

The problem of partitioning an unstructured mesh has attracted the imaginations of many workers for more than twenty years. A *partitioner* will decompose a computational grid into partitions (subdomains). To ensure efficient execution, assuming an efficient implementation, the produced subdomains need to have the following characteristics:

- minimum number of interface nodes between two adjacent partitions; therefore reducing the number of nodes needed to be exchanged between the processors which have those nodes
- each processor has the approximately equal number of data items; this ensures that each processor has to do the same amount of work

For many meshes it can be computationally prohibitive to find an optimal partition and computationally expensive to find a near optimal partition. On the other hand a reasonable partition may be calculated with little effort. What is required of a mesh partitioning algorithm is a high quality of partition at a low cost. The time required to calculate the partition must be insignificant in proportion to the time for the irregular code to execute. High quality means a balanced load, short interfaces and a small number of interfaces.

Mesh points are partitioned to minimize communication. Some promising, recent partitioning heuristics use one or several of these types of information: the spatial locations of mesh vertices, the connectivity of the vertices, and an estimate of the computational load associated with each mesh point. For instance, a developer might choose a partitioner that is based on coordinates. A coordinate bisection partitioner decomposes data using the spatial locations of mesh vertices. If the developer chooses a graph–based partitioner, the connectivity of the mesh could be used to decompose the mesh.

A fundamental objective in finding a partition is to balance the computational effort or load required in each sub–domain. The simplest approach is to assume that the load per element is homogeneous throughout the mesh. In this case the partition should have as near equal numbers of elements per partition as possible. Should the load be inhomogeneous then a weight or cost function may be applied to the elements to achieve a cost balanced partition. An important consideration in load balancing is that it is not so much essential to achieve a totally uniform balance of load but rather that no processor should have significantly more than average load. Any processor with an exceptional work load will cause all other processors to incur idle time with resultingly poor parallel performance. The next step in parallelizing this application involves assigning equal amounts of work to processors. For example, an unstructured Euler solver (see code in Fig. 4.46) consists of a sequence of loops that sweep over a mesh. Computational work associated with each loop must be partitioned to balance the load. Therefore, mesh edges are partitioned so that load balance is maintained and computations employ mostly locally stored data. Applying a weight to the nodes (perhaps based upon the number of connected elements and/or some other parameters) and then partitioning the weighted list can give an improved balance.

Complex aerodynamic shapes require high–resolution meshes and, consequently, large numbers of mesh points. Therefore, ideally, the partitioning of the mesh should be carried out at run time in parallel. As the number of processors NP and the number of mesh points N increase, an $O(N)$ partitioning algorithm may become unacceptable for a solver running at $O(f(N)/NP)$ [186]. Therefore, the utilization of highly efficient parallel algorithm is very urgent. However, few of the available partitioning algorithms are suitable for parallel implementation. Moreover, currently, even sequential partitioners must be coupled to user programs manually. This manual coupling is particularly troublesome and tedious when a user wishes to make use of parallelized partitioners. Furthermore, partitioners use different data structures and are very problem dependent, making it extremely difficult to adapt to different (but similar) problems and systems. It is reasonable to define common data structures, which can be used by any general data partitioner. These interface data structures are to be generated by the compiler automatically due to the high level partitioning specification provided by the user. The data structures enable linking partitioners to programs. They store information on which data partitioning is to be based; they can utilize spatial, load, or connectivity information or combination of them. For example, a code segment can be inserted into the SPMD program to construct a graph–structure representing the access patterns of the distributed arrays in the loop. The graph is constructed by executing a modified version of the loop which forms a list of edges. Note that the operations in the loop are not performed. The generated graph is passed to a graph partitioner which returns the new array distribution in the form of a translation table (see definition 4.3.2, for example.

Parallel partitioners are provided by a runtime library which may also include description of each partitioner; this information may be utilized by the compiler for checking the user specification.

In our design and implementation, we have used the partitioners and other supporting procedures (e.g., procedures for generating interface data structures for partitioners) from the CHAOS runtime library [217]. These features are briefly described below.

4.4.2 Partitioners of the CHAOS Runtime Library

CHAOS runtime library can be used either by application programmers to parallelize irregular problems, or can be used by parallelizing compilers to generate code for irregular problems. It provides support for inspector and executor codes, partitioning data and work, and data redistribution. In order to achieve good performance, it offers parallel versions of those heuristic methods that have been developed for sequential partitioning.

Currently, the CHAOS parallel partitioners implement the following partitioning methods:

1. Recursive Coordinate Bisection (RCB) [31]
2. Recursive Inertial Bisection (RIB) [205]
3. Weighted Recursive Inertial Bisection (WRIB)
4. Chain Partitioning (CP) [35, 199]
5. Recursive Spectral Bisection (RSB) [19]
6. Bin Packing (BP)

Partitioners from the first four classes partition data based on spatial information and/or computational loads. The RSB partitioner finds partitioning by connectivity of graphs; BP considers only computational loads. For example, RCB is a simple geometric scheme in which the grid points of the mesh are sorted into order along one axis (normally the longest) and then bisected. This process is repeated recursively on each partition until the required number of partitions is obtained. A variant of the scheme is the Orthogonal Coordinate Bisection (OCB) in which the sort axis is alternated at each recursion.

The above set of partitioners can be extended by other partitioners, for example, by a partitioner based on the Simulated Annealing (SA) optimization method that borrows ideas from a statistical mechanics approach to annealing in a cooling solid [251].

The set of parallel partitioners provided by CHAOS has been fully integrated into VFCS.

4.4.3 High Level Language Interface

In this subsection, we show how a higher level language interface may be provided in which control over the data and work distribution in a FORALL loop can be delegated to a combination of a compiler and runtime system.

Recall that irregular computations in Vienna Fortran can be specified by explicit data parallel loops – forall loops. The general form of a forall loop which allows to specify a partitioner is the following:

FORALL $(I_1 = sec_1, \dots, I_n = sec_n)$ [$Work\_distr\_spec \mid Partitioner\_spec$]
  $Loop\text{-}body$
END FORALL

Partitioner_spec denotes a specification of the strategy to be automatically applied by the programming environment to the forall loop for calculating optimal data and work distributions. In this context, *Work_distr_spec* only specifies the initial work distribution used for parallel computation of information that will be used for data partitioning. This initial work distribution will usually be replaced by the work distribution performed after data partitioning.

Work_distr_spec and *Partitioner_spec* can be either introduced in the source program or specified interactively during the parallelization process.

(a) Fortran code

```
PARAMETER :: N1 = ..., N2 = ...
REAL, DIMENSION (N1) :: X, Y, Z
INTEGER, DIMENSION (N2) :: ID
...
DO i = 1, N2
    Y(ID(i)) = Y(ID(i)) + X(ID(i)) * Z(ID(i)))
END DO
```

(b) Vienna Fortran code

```
PARAMETER :: NP = ..., N1 = ..., N2 = ...
PROCESSORS :: P(NP)
REAL, DIMENSION (N1), DYNAMIC, DISTRIBUTED (BLOCK) :: X, Y, Z
INTEGER, DIMENSION (N2), DISTRIBUTED (BLOCK) :: ID
...
FORALL i = 1, N2 USE (RSB_PART(), ARR = (X, Y, Z))
    REDUCE ( ADD, Y(ID(i)), X(ID(i)) + Z(ID(i)))
END FORALL
```

Fig. 4.44. Specification of a Partitioner

A partitioner specification is illustrated by a simple Vienna Fortran code fragment in Fig. 4.44. Using the *use-clause* enables the programmer to select a partitioning algorithm from those provided in the environment. The parallel partitioner will find a new irregular distribution for data arrays specified in the *use-clause* and these data arrays will be redistributed accordingly. In the second step a new work distribution will be determined for the loop to minimize load imbalances and maximize data locality. Information needed for a partitioner may be passed by means of arguments. Data partitioners make use of different kinds of information. Typically, they operate on data structures that represent a combination of the graph connectivity, geometrical and/or weight (computational load) information, and return an irregular assignment of data elements to processors in the form of the mapping array. In our example, the recursive spectral bisection (RSB) partitioner is applied to the FORALL loop. As we will see later, additional information (e.g., load or geometry information) used in the partitioning process can be passed to the partitioner. In our example, a new distribution is computed for arrays

X, Y, and Z which will retain that distribution after the loop has completed the execution. By default, a new data distribution is computed for all data arrays, if no ARR specification is introduced. The interface data structure for RSB is constructed using the indirection array access pattern $ID(i)$. Additional actions to be done for the FORALL loop may be defined through a set of options. For example, it may be sometimes desirable to restore the distribution of one or more of the newly partitioned arrays after the loop has been executed.

The interface to the partitioners is generated by VFCS automatically.

4.4.4 Interactive Specification of the Partitioning Strategy

VFCS provides the user with the opportunity to specify the appropriate partitioners and their arguments for a set of selected loops interactively. The system immediately checks the correctness of the specification, stores the specification in the internal representation of the program, and displays it at the do loop header. The snapshots in Fig. 4.45 illustrate a selection of the partitioner. To specify a partitioner the user has to mark the target forall loop and select the menu [Parallelization : Partitioner]. The system then provides a list of partitioners currently available in the environment (see the upper snapshot). Then the user clicks, for example, on the [Coordinate] item to select the RCB partitioner. The system asks to pass the information about dimensionality and spatial location of array elements. The user introduces arrays (for example, X, Y and Z) as objects storing the coordinates (see the lower snapshot).

4.4.5 Implementation Issues

The partitioning process involves three major steps:

1. Data Distribution.
   This phase calculates how data arrays are to be decomposed by making use of the selected partitioner. In the case of the RSB partitioner, the compiler generates the coupling code which produces a distributed data structure called *Runtime Data Graph* (RDG). This graph is constructed from array access patterns occurring in the forall loop. The RDG structure is then passed as input to the RSB partitioner. RCB, WRCB, RIB and WRIB partitioners are passed the spatial location of the array elements and/or information associating the array elements with computational load in the form of data arrays. The output of the partitioners is the new data distribution descriptor which is stored as a CHAOS construct called *translation table* representing a part of the distribution descriptor.
2. Work Distribution.
   Once the new data distribution is obtained, the loop iterations must be distributed among the processors to minimize non–local array references.

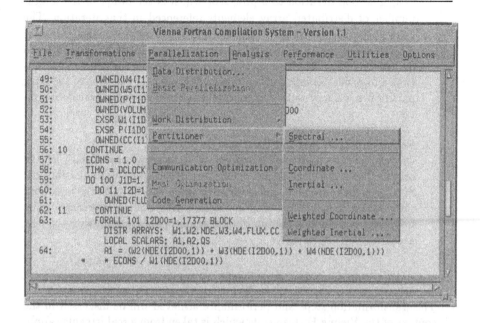

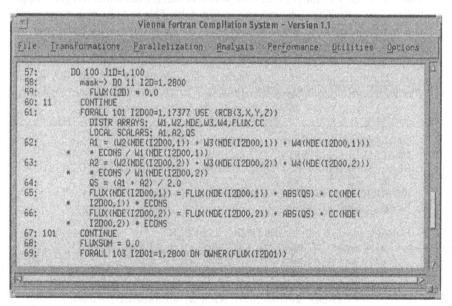

Fig. 4.45. Selecting a Partitioner

For work partitioning the *almost owner computes* paradigm is chosen that assigns a loop iteration to a processor which is the owner of the largest number of distributed array elements referenced in that iteration. The work partitioner operates on the runtime data structure, called *Runtime Iteration Graph* (RIG), which lists for each iteration i all indices of each distributed array accessed in this iteration. RIG is constructed at run-time and passed as input to the work partitioner. The output of the work partitioner is a new work distribution descriptor stored as *translation table*.

3. **Redistribution of Data and Iterations.**
   To complete the partitioning process, all data arrays associated with the original block distribution are redistributed to the new data distribution. Similarly, the loop iterations are redistributed according to the new work distribution specification. The indirection arrays (data references) must be redistributed following the work distribution to conform with the new execution set.

4.4.6 Illustration of the Approach

The transformation steps and performance achieved will be discussed in the context of the Vienna Fortran code which is taken from a real irregular application program. The code is shown in Fig. 4.46. This code fragment represents the kernel loop from the unstructured 3–D Euler solver (EUL3D) and will be used as a running example to illustrate our parallelization methods. The loop represents the sweep over the edges, the computationally most intensive part. The indirection array *nde* and the data arrays *w1, w2, w3, w4, cc*, and *flux* have initially got block distributions.

```
PARAMETER :: NNODE = ..., NEDGE = ..., NP = ...
PROCESSORS pp1(NP)
REAL, DIMENSION(NNODE), DYNAMIC, DISTRIBUTED(BLOCK) ::    &
     flux, w1, w2, w3, w4, cc
INTEGER, DISTRIBUTED(BLOCK,:) :: nde(nedge,2)
REAL :: econs, a1, a2, qs
...
FORALL i = 1, NEDGE DISTRIBUTE(BLOCK)
     a1 = econs*(w2(nde(i,1)) + w3(nde(i,1)) + w4(nde(i,1))) / w1(nde(i,1))
     a2 = econs*(w2(nde(i,2)) + w3(nde(i,2)) + w4(nde(i,2))) / w1(nde(i,2))
     qs = (a1 + a2) / 2.0
     REDUCE(ADD, flux(nde(i,1)), ABS(qs) + cc(nde(i,1))*econs)
     REDUCE(ADD, flux(nde(i,2)), ABS(qs) + cc(nde(i,2))*econs)
END FORALL
```

Fig. 4.46. Kernel Loop of the Unstructured Euler Solver EUL3D

The partitioning code generated by the VFCS for the EUL3D kernel loop given in Fig. 4.46 is depicted in Fig. 4.47, 4.48 and 4.49, and the inspector/executor code in Fig. 4.50.

Data Partitioning

Fig. 4.47 depicts the data partitioning codes generated for the RSB partitioner (the upper part) and the RCB partitioner (the lower part). Both partitioners return an irregular assignment of array elements to processors in the form of the translation table.

To implement automatic data distribution a distributed data structure, called *Runtime Data Graph* (RDG) must be constructed. Input data for the construction of RDG are computed by the parallel execution of a modified version of the FORALL loop. Therefore, loop iterations of the loop are initially distributed among processors according to the initial work distribution specification. If no initial work distribution is specified, some uniform work distribution is chosen. In our implementation, we use the block distribution which is computed by the primitive build_ES_block. It accepts the lower and upper bounds of the iteration space (*1, nedge*) and returns the execution set (*exec*) and its size (*exec_size*). The data arrays that will be partitioned are also initially distributed in block fashion. The local index set (*lis*) and its size (*lis_size*) are computed from the lower and upper bounds (*1, nnode*) of the array dimensions that are to be distributed by the primitive build_LIS_block.

The RDG is generated by adding an undirected edge between node pairs representing the left hand side and the right hand side array indices for each loop iteration.

First, the procedure elim_dup_edges generates a local RDG on each processor using the distributed array access patterns that appear in the iterations involved in *exec*. This procedure is called once for each statement that accesses distributed arrays on both the left hand side and the right hand side of the statement. The right hand side access pattern and left hand side access patterns are recorded for every statement separately through a do loop, and are passed as two integer arrays to the elim_dup_edges. Procedure elim_dup_edges uses a hash table identified by a hashindex *ght* to store unique dependency edges.

In the next step, the local graphs are merged to form a distributed graph. Procedure generate_RDG is called on each processor to combine the local adjacency lists in the hash table into a complete graph (closely related to compressed sparse row format). The RDG is distributed so that each processor stores the adjacency lists for a subset of the array elements. Edges are grouped for each array element together and then stored in the list represented by the integer array *csrc*. A list of pointers, represented by the integer array *csrp*, identifies the beginning of the edge list for each array element in *csrc*.

```
      TYPE (DIST_DESC), POINTER :: ddd, idd
      TYPE (TRA_TAB), POINTER :: dtrat, itrat
      INTEGER, DIMENSION (:), POINTER :: lis
      INTEGER, DIMENSION (:), ALLOCATABLE :: exec
      ...
C  Constructing local index set (initial block distribution)
      CALL build_LIS_block (1, nnode, lis, lis_size)

C  Constructing the execution set (initial block distribution)
      CALL build_ES_block (1, nedge, exec, exec_size)
```

```
C  DATA PARTITIONING using RSB

C   Initialize hash table to store RDG
      ght = Init_Hash_Tab (exec_size)

C   Constructing local RDG
      n1 = 1; n2 = 1
      DO i=1, exec_size
          globref1(n1) = nde(i,1)
          n1 = n1 + 1
          globref2(n2) = nde(i,1)
          globref2(n2+1) = nde(i,2)
          n2 = n2 + 2
      END DO
      CALL elim_dup_edges (ght, globref1, n1-1, globref2, n2-1)
      n1 = 1; n2 = 1
      DO i=1, exec_size
          globref1(n1) = nde(i,2)
          n1 = n1 + 1
          globref2(n2) = nde(i,2)
          globref2(n2+1) = nde(i,1)
          n2 = n2 + 2
      END DO
      CALL elim_dup_edges (ght, globref1, n1-1, globref2, n2-1)

C  Generating distributed RDG: stored in structures csrp and csrc
      CALL generate_RDG (ght, lis, lis_size, csrp, csrc)

C  RSB partitioner: returns translation table pointer dtrat
      CALL partitioner_RSB (lis, lis_size, csrp, csrc, exec_size, dtrat)
```

```
C  DATA PARTITIONING using RCB

C   RCB partitioner: returns translation table pointer dtrat; x, y, z are coordinate arrays
      CALL partitioner_RCB (x, y, z, lis_size, dtrat)
```

Fig. 4.47. Data Partitioning for the EUL3D Kernel Loop

The RDG representation (arrays *csrc* and *csrp*) of the distributed array access patterns is passed to the parallel partitioner partitioner_RSB which returns a pointer to the translation table *dtrat* describing the new array distribution.

The lower part of Fig. 4.47 illustrates a call to the RCB partitioner. Here, the use of three–dimensional geometric information is shown. Arrays x, y, and z carry the spatial coordinates for elements in the data arrays. Note that the size of the local index set (*lis_size*) was computed through the primitive build_LIS_block.

Iteration Partitioning

The newly specified data array distribution are used to decide how loop iterations are to be distributed among processors. This calculation takes into account the processor assignment of the distributed array elements accessed in each iteration. A loop iteration is assigned to the processor that owns the maximum number of distributed array elements referenced in that iteration.

```
C   ITERATION PARTITIONING

C   Collecting data array accesses
      n = 1
      DO i=1, exec_size
         globref(n) = nde(i,1)
         globref(n+1) = nde(i,2)
         n = n + 2
      END DO

C   Iteration partitioner: returns translation table pointer itrat
      CALL partitioner_Iter (dtrat, globref, exec_size, n-1, itrat)
```

Fig. 4.48. Iteration Partitioning for the EUL3D Kernel Loop

Distributing loop iterations is done via procedure partitioner_Iter which includes the generating the runtime data structure, called *Runtime Iteration Processor Assignment* graph (RIPA) and call to the iteration distribution procedure. The RIPA lists for each iteration, the number of distinct data references associated with each processor. It is constructed from the list of distinct distributed array access patterns encountered in each loop iteration, which are recorded in an integer array and passed as a parameter to the partitioner_Iter procedure. Using the RIPA the iteration distribution subroutine assigns iterations to processors. The new loop iteration distribution is output in form of the translation table *itrat*.

Redistribution of Arrays

Once the new distributions have been specified data and indirection arrays must redistributed based on it. This is done in two steps which are shown in Fig. 4.49.

C **REDISTRIBUTING Arrays**

C Constructing schedules for redistribution
      CALL Remap (dtrat, lis, dsched, newlis, newlis_size)
      CALL Remap (itrat, exec, isched, newexec, newexec_size)

C Redistributing data arrays
      CALL Gather (flux(1), flux(1), dsched)
      CALL Gather (w1(1), w1(1), dsched)
      CALL Gather (w2(1), w2(1), dsched)
      CALL Gather (w3(1), w3(1), dsched)
      CALL Gather (w4(1), w4(1), dsched)
      CALL Gather (cc(1), cc(1), dsched)
      CALL Free_sched (dsched)

C Redistributing indirection array
      CALL Gather (nde(1,1), nde(1,1), isched)
      CALL Gather (nde(1,2), nde(1,2), isched)
      CALL Free_sched (isched)

C Updating runtime distribution descriptors
      CALL Update_DD (ddd, newlis, newlis_size, dtrat)
      CALL Update_DD (idd, newexec, newexec_size, itrat)

Fig. 4.49. Redistribution of Arrays for the EUL3D Kernel Loop

1. The procedure **Remap** is called which returns a schedule pointer *dsched* and a new local index set *newlis* and its size *newlis_size* corresponding to the new distribution of the data arrays; and similarly, a schedule pointer *isched* and the new list of loop iterations *newexec* and its size *newexec_size* corresponding to the new work distribution.

2. The schedule pointer *dsched* is passed to the CHAOS routine **Gather** to redistribute data arrays among processors.

   The data references in each iteration must be redistributed to conform with the new loop iteration distribution. That is, the indirection array elements *nde(i,1)* and *nde(i,2)* are moved to the processor which executes the iteration *i*. The communication schedule *isched* is passed to the CHAOS routine **Gather** to redistribute the indirection array *nde*.

   After redistributing arrays the procedure **Update_DD** is called to update the runtime distribution descriptors for data arrays *ddd*, and for indirection arrays *idd*, according to the new irregular distributions.

Inspector and Executor

Inspector and executor for the EULER3D Kernel Loop are shown in Fig. 4.50.

4.4.7 Performance Results

This section presents the performance of the automatically parallelized 3–D Euler solver (EUL3D), whose kernel is introduced in Section 4.4.3. We examined the code included in the outer loop with 100 iteration steps, using the input datasets with NNODE = 2800 and NEDGE = 17377. Two versions of the code have been elaborated to evaluate the influence of different kinds of data distributions on parallel–program execution time. In the first version, all data arrays are block distributed, in the second version, data arrays accessed in irregular code parts are initially declared as block distributed, and after applying a partitioner they are redistributed irregularly according to the calculated new mapping. We introduce only times for RSB and RCB partitioners. Times for other partitioners (RIB, WRCB and WRIB) are like the times for the RCB partitioner. Experiments were performed on Intel iPSC/860 system, the sequential code on 1 processor, the parallel code on 4, 8, and 16 processors. The generated parallel code was optimized in such a way that the partitioning code and the invariant part of the inspector was moved out of the outer loop, since the communication patterns do not change between the solver iterations. The timing results (in seconds) are summarized in Table 4.3.

Table 4.3. Performance Results of the EUL3D Solver

Sequential code					
35.37					
Parallel code					
Computational Phases	Irregular Distribution				
	RSB			RCB	
	Number of Processors				
	4	8	16	4	8	16
Data Partitioning	3.20	10.92	12.47	0.17	0.18	0.33
Work Partitioning	0.59	0.40	0.30	0.59	0.35	0.25
Total	20.31	19.56	17.14	15.64	8.23	4.91
	Block Distribution				
	Number of Processors				
	4		8		16	
Total	25.84		18.06		13.14	

C **INSPECTOR Code**

C Constructing global reference list for data arrays;
C Indirection array conforms with the new execution set
```
      n = 1
      DO  k=1, newexec_size
          globref(n) = nde(k,1)
          globref(n+1) = nde(k,2)
          n = n + 2
      END DO
```

C Computing schedule and local reference list for data arrays
```
      CALL  Localize (ddd, n-1, globref, locref, nonloc, dsched)
```

C **EXECUTOR Code**

C Initialize buffer area of flux (to implement reduction operations)
```
      DO  k=1, nonloc
          flux(newlis_size + k) = 0.0
      END DO
```

C Gather non−local elements of cc and w1, w2, w3, w4
```
      CALL  Gather (cc(newlis_size+1), cc(1), dsched)
      CALL  Gather (w1(newlis_size+1), w1(1), dsched)
      CALL  Gather (w2(newlis_size+1), w2(1), dsched)
      CALL  Gather (w3(newlis_size+1), w3(1), dsched)
      CALL  Gather (w4(newlis_size+1), w4(1), dsched)
```

C Transformed loop body
```
      n = 1
      DO  k=1, newexec_size
          a1 = econs * (w2(locref(n)) + w3(locref(n)) +            &
                   w4(locref(n))) / w1(locref(n))
          a2 = econs * (w2(locref(n+1)) + w3(locref(n+1)) +        &
                   w4(locref(n+1))) / w1(locref(n+1))
          qs = (a1 + a2) / 2.0
          flux(locref(n)) = flux(locref(n)) + ABS(qs) +           &
                   cc(locref(n)) * econs
          flux(locref(n+1)) = flux(locref(n+1)) + ABS(qs) +       &
                   cc(locref(n+1)) * econs
          n = n + 2
      END DO
```

C Accumulate values of non−local elements of flux
```
      CALL  Scatter_add (flux(newlis_size+1), flux(1), dsched)
```

Fig. 4.50. Inspector and Executor for the EUL3D Kernel Loop

BIBLIOGRAPHICAL NOTES

Strategies for mapping unstructured mesh based computational mechanics codes for distributed memory systems are developed in [186]. In particular, this work examines the method of parallelization by geometric domain decomposition. Chapman et al. [73] propose the *use–clause* for the HPF INDEPENDENT loop which enables the programmer to specify a partitioner. Ponnusamy et al. [215] propose a directive that can direct a compiler to generate a data structure on which data partitioning is to be based and a directive for linking this data structure to a partitioner. An experimental Fortran 90D implementation is able to couple the CHAOS partioners to forall statements. Ponnusamy [214] describes the coupling of CHAOS partitioners to SPMD codes. In [64], we describe the utilization of an external partitioner. The coupling of CHAOS partitioners to the VFCS is described in more detail in [52].

4.5 Compile Time Optimizations for Irregular Codes

4.5.1 Introduction

This section describes the design of compile time optimizations for irregular applications. It focuses on

- the reduction of the runtime overhead caused by the precomputation needed to
  - derive work distribution
  - generate communication schedules and carry out index conversion
- communication optimization
- the reduction of the memory overhead

Optimizations are based on high level specifiers introduced in the internal program representation and the program analysis and transformation methods developed in the field of Partial Redundancy Elimination. The specifiers denote computations whose redundancy or partial redundancy elimination is the objective of the optimization. These methods can also be directly applied to other optimizations in irregular applications, in particular, to the reduction of the runtime and memory overheads caused by the precomputations needed to initialize data distribution descriptors and dynamic data and work partitioning.

Motivation and Objectives

In Section 4.3, we described the strategy for parallelization and execution of irregular codes. According to this strategy, the compiler generates a code which constructs a runtime representation for distributed arrays, referred to as *distribution descriptors* (DDs), and generates three code phases for *each* irregular construct, called the work distributor, the inspector, and the executor. Optionally, a partitioning phase can be inserted between the work distributor and the inspector, as presented in Section 4.4. All the phases are supported by the appropriate runtime library, for example, the CHAOS library. The functionality of these phases can be characterized as follows:

- The work distributor determines how to spread the work (iterations) among the available processors. In each processor, the work distributor computes the *execution set*, i.e. the set of loop iterations to be executed in this processor. This set is stored in a (processor–local) data structure called the *Execution Set Descriptor* (ESD).

- The data and work partitioner computes the data and work distribution according to the specified criteria. The partitioner generates a (processor–local) runtime descriptor called the *Data and Work Distribution Descriptor* (DWDD) which includes data distribution descriptors for arrays occurring in the FORALL loop and the execution set descriptor for that loop.

- The inspector analyzes the communication patterns of the loop, computes the description of the communication, and derives translation functions between global and local accesses. The communication needed for an array object is described by a data structure called *schedule*. The result of each global to local index conversion is stored in a list of local indices which will be referred to as *local look–up table* (*llut*) in the following text. The pair (*schedule, llut*) is called the *data availability descriptor* (DAD).

- The executor performs the actual communication described by schedules and executes the loop iterations included in the execution set. Addressing of distributed objects is done by means of local look–up tables.

   Execution of the loop iterations performed in the execution phase corresponds to the computation specified in the original program. On the other hand, the distributor, partitioners, inspector, and communication performed in the executor phase introduce additional runtime and memory overhead associated with the parallelization method. Reduction of this overhead is a crucial issue in the compiler development. Sophisticated compile time representation and analysis techniques have to be applied to find out cases when the reuse of results of the data distribution descriptor construction, the work distributor, the partitioner, the inspector, and communication is safe and when it is possible to free memory occupied by large data structures that implement the descriptors. In this section, we specify techniques for performing the above mentioned optimizations. The key idea underlying these techniques is to do the code placement so that *redundancies are removed* or *reduced*. These techniques are based upon a data flow framework called Partial Redundancy Elimination.

The rest of this section is organized as follows. Section 4.5.2 discusses the basic partial redundancy elimination framework. The features of our approach are outlined in Sections 4.5.3, 4.5.4 and 4.5.5. Section 4.5.6 deals in detail with intraprocedural optimizations. Interprocedural optimizations are only touched in Section 4.5.7.

4.5.2 Partial Redundancy Elimination

Partial Redundancy Elimination (PRE) unifies the traditional optimizations of code movement, common subexpression elimination, and loop optimization which are described, for example, in [8, 142]. PRE has been originally proposed by Morel and Renvoise [192] and has been improved considerably by Dhamdhere et al. [102, 104, 103]. Based on the Morel/Renvoise approach, new frameworks for the elimination of partially dead code, optimal assignment placement, and excluding any unnecessary code motion have been presented by Knoop et al. [161, 162, 164]. The last works also handle second order effects due to the mutual dependences between transformation steps (e.g., between expression and assignment hoisting). The functionality of PRE can be summarized as follows.

The program is modeled by a directed flow graph called the *control flow graph* (CFG) the nodes of which are basic blocks. A *basic block* is a sequence of consecutive statements in the program text without any intervening procedure calls or return statements, and no branching except at the beginning and at the end [8]. The CFG is assumed to be single–entry/single–exit, i.e. it has two dedicated nodes *entry* and *exit*. A path in CFG is a sequence of nodes $(b_1, b_2, ..., b_k)$, where for all $1 \leq i < k$, b_i is connected with b_{i+1} by an edge $b_i \rightarrow b_{i+1}$.

Consider any computation of an expression. In the program text, we may want to optimize its placement, i.e. to place this computation so that the result of the computation is used as often as possible, and redundant computations are removed. For convenience, we refer to any such computation whose placement we want to optimize as a *candidate*. When an expression is included in an assignment statement, optimization of placement of this expression is associated with the statement placement. In this context, we focus on assignment statements which are of the form:

$$u \leftarrow E(o_1, ..., o_n)$$

where u is a variable (introduced by the programmer or parallelizing compiler), and E is an expression using other variables $o_1, ..., o_n$ (*operands* of E). It is assumed that all such kind of assignments are transformed into a canonical form in which an assignment can be either an *expression evaluation* or a *single value assignment*.

An expression evaluation is of the form:

$$e \leftarrow E(o_1, ..., o_n)$$

where e is an *expression temporary* (see explanation below). Variables $o_1, ..., o_n$ are called *influencers* of this expression evaluation.

A single value assignment is of the form:

$$u \leftarrow e$$

where e is an expression temporary.

The expression temporaries are artificial objects and rather restricted as compared to user variables. Each expression temporary

- is uniquely associated with an expression pattern
- will only be defined in evaluations of the corresponding expression pattern
- serves exclusively as an intermediate storage from a program point d where the expression is evaluated, to the point u where the resulting value is used. No modifications to operands are allowed to take place on any CFG path from d to u, or in other words, in–place evaluation of E just before u would yield the same value – a property which will turn out to be important subsequently.

These characteristics are used to detect, and later eliminate *partial redundancies*. This is illustrated in the setting depicted in Fig. 4.51 which shows:

(a) a trivial case of a (total) redundancy: The second evaluation of E is redundant because it will yield the same value as the first one (no operands have been modified between them).
(b) a partial redundancy: The second evaluation of E is redundant if control reaches $b5$ using the left branch, but not redundant when the right branch is used. Here another evaluation of E can be inserted in $b3$, making the one in $b5$ totally redundant, at which point it can be deleted.
(c) the loop–invariant code motion, a special case of PRE: Consider block $b2$. If control reaches this block via the loop back edge (i.e. in the second and all further iterations), E is already available here. However this is not the case when control reaches $b2$ coming from $b1$; thus the evaluation in $b2$ is only partially redundant. If, however, an additional evaluation of E is inserted in $b1$, the one in $b2$ again becomes totally redundant and can be

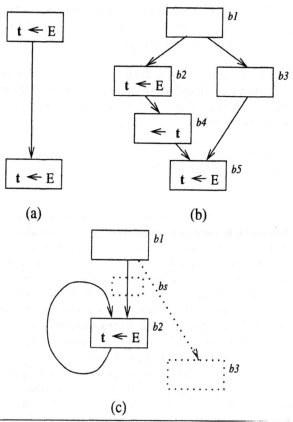

Fig. 4.51. Partial redundancies

deleted – which in effect means that it has been hoisted out of the loop. Note that insertion in $b1$ is not possible for safety reasons if $b1$ lies on another path $b1 \rightarrow b3$ which originally did not contain an evaluation of E. In that case, a synthetic block bs has to be inserted between $b1$ and $b3$ in which the evaluation of E can then be safely inserted. The edge $b1 \rightarrow b2$ is called *critical* in that case.

In the following we now summarize the key data flow properties that are computed as part of the PRE framework[15].

Availability. Availability of a candidate C at any point p in the program means that C lies at each of the paths leading to point p and if C were to be placed at point p, C would have the same result as the result of the last occurrence on any of the paths.

---

[15] An excellent overview of the PRE approach can be found in [3].

Partial Availability. Partial availability of a candidate C at a point p in the program means that C is currently placed on at least one control flow path leading to p and if C were placed at the point p, C would have the same result of the last occurrence on at least one of the paths.

Anticipability. Anticipability of a candidate C at a point p in the program means that C is currently placed at all the paths leading from point p, and if C were to be placed at point p, C would have the same result as the first occurrence on any of the paths.

Transparency. Transparency of a basic block with respect to a candidate means that none of the influencers of the candidate are modified in the basic block.

If a candidate is placed at a point p in the program and if it is (partially) available at the point p, then the occurrence of the candidate at the point p is (partially) redundant. Anticipability of a computation is used for determining whether the placement will be *safe*. A safe placement means that at least one occurrence of the candidate will be made redundant by this new placement (and will consequently be deleted). Performing safe placements guarantees that along any path, number of computations of the candidate are not increased after applying optimizing transformations.

By solving data flow equations on the CFG of a procedure, the properties Availability, Partial Availability and Anticipability are computed at the beginning and end of each basic block in the procedure. Transparency is used for propagating these properties, e.g. if a candidate is available at the beginning of a basic block and if the basic block is transparent with respect to this candidate, then the candidate will be available at the end of the basic block.

Based on the above data flow properties, another round of data flow analysis is done to determine whether it is possible to place computation at the beginning or end of each basic block. These results are then used for determining final placement and deletion of the candidates. We use the above discussed features of PRE for explaining several new optimizations later in this section.

4.5.3 Specifiers of the Candidates for Placement

In the initial parallelization phase, instead of direct generation of parallel code as it was presented in Sections 4.3 and 4.4, high–level information structures called *specifiers of candidates for placement* are constructed in the internal representation (IR). These serve as a basis for the compile time analysis, optimizing transformations, and target code generation.

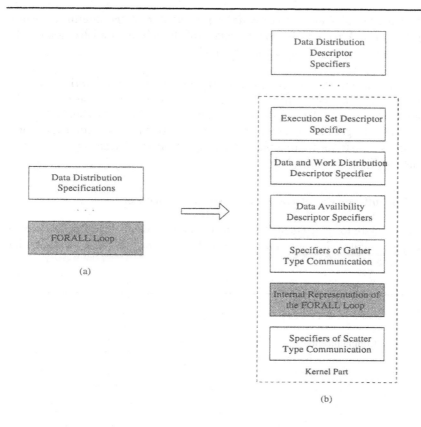

Fig. 4.52. Inserting Specifiers

The initial parallelization phase is carried out in two major steps. First, the distribution and alignment specifications of the input program are processed and specifiers of the computation of data distribution descriptors are inserted into the IR. In the second step, the IR is extended by data structures which represent the compile time knowledge about forall loops. In the extended IR, each forall loop is condensed to a single node which encompasses the pointers to the following components:

- IR of the forall loop
- specifiers for the computation of
  - the execution set descriptor (ESD)
  - data availability descriptors (DADs); one DAD for each distributed array accessed in the loop
  - data and work distribution descriptor (DWDD); one DWDD for each forall loop to which a data and work partitioner has to be coupled
- specifiers of the gather or scatter type communication

In our examples we will use a *flat representation* of the intermediate program form. According to this, the effect of the initial parallelization can be expressed in a way as shown in Fig. 4.52.

The optimization phase traverses IR, examines the information provided by specifiers and performs the appropriate program transformations. Via the analysis, the compiler can detect redundancies and partial redundancies.

In the target code generation phase, the internal representation of the program is transformed into a parallel Message Passing Fortran target program; the specifiers are expanded.

4.5.4 Influencers of the Candidate for Placement

A candidate for placement[16] depends on a set of agents called *influencers* of this candidate. Computation of each influencer may also be a candidate for placement.

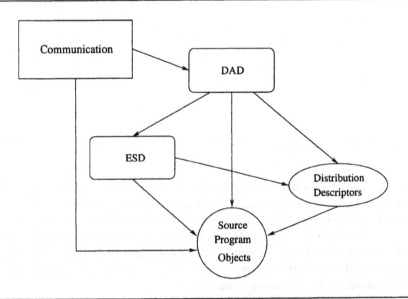

Fig. 4.53. Relation Candidate–Influencers Considered in Communication Placement

For example, consider communication of GATHER type as a candidate for placement. We call this candidate as the *main candidate*. For this case,

---

[16] Recall that we refer to any computation or communication whose placement we want to optimize as a *candidate* for placement.

relation Candidate–Influencers is graphically sketched in Fig. 4.53. It is important to be aware of this relationship when developing the appropriate data flow framework. This framework should be able to optimize not only the placement of main candidates but of their influencers as well. The safety criteria is that the influencer is placed in such a way that its value is always computed before the agent at a lower hierarchical level that is influenced by that influencer is evaluated.

4.5.5 The Form of Specifiers

Each specifier has the form:

$$\boxed{\text{result} := \text{Operator (influencers)}}$$

with

– *Operator*, the name denoting the candidate for placement; it may be communication or computation
– *result*, the name denoting a data structure that stores and describes the result of the *Operator* application
– *influencers*, a list of the influencers of the candidate.

In the following, we only deal with the concrete form of specifiers that describe the computation of ESD and DAD descriptors, and communication. The form of specifiers that describe the computation of DD and DWDD descriptors is introduced in [61].

Execution Set Descriptor Specifier

Relation Candidate–Influencers which must be considered by the placement of an execution set descriptor specifier is depicted in Fig. 4.54

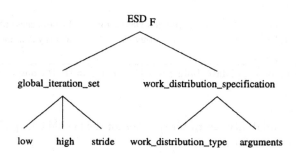

Fig. 4.54. Relation Candidate–Influencers Considered in ESD Computation Placement

The execution set descriptor specifier for the FORALL loop F has the following form:

$$\boxed{\text{ESD}_F := \text{Compute\_ESD (low, high, stride, wd\_type, wd\_arguments)}}$$

with

- ESD_F, execution set descriptor
- *low, high, stride*, description of the global iteration set taken from the forall loop header
- *wd_type*, work distribution type (e.g., *on_owner*)
- *wd_arguments*, arguments specific for *wd_type* (e.g., the data distribution descriptor and access pattern for the *on_owner* work distribution type)

For each type of work distribution, different work distribution arguments are involved in the specifier. Regular work distribution types, BLOCK and CYCLIC, need no arguments. For the work distribution of *on_owner* type, for example, the distribution descriptor of the associated array and the description of its access pattern are involved in the specifier.

Data Availability Descriptor Specifier

The data availability descriptor is constructed for each distributed array which is indirectly accessed in the given forall loop. Relation Candidate–Influencers which must be considered in the placement of a DAD specifier is depicted in Fig. 4.55.

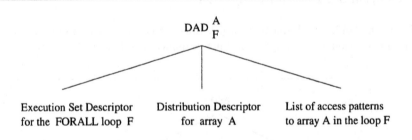

$$DAD^A_F$$

Execution Set Descriptor   Distribution Descriptor   List of access patterns
for the FORALL loop F   for array A   to array A in the loop F

Fig. 4.55. Relation Candidate–Influencers Considered in DAD Computation Placement

The data availability descriptor specifier for the FORALL loop F and array A has the following form:

$$\boxed{\text{DAD}^A_F := \text{Comp\_DAD } (\text{ESD}_F, \text{DD}^A, \text{access\_patterns}^A_F)}$$

with

- DAD_F^A, data availability descriptor for array A in the **FORALL** loop F
- ESD_F, execution set descriptor for the **FORALL** loop F
- DD^A, distribution descriptor of array A
- $access\_patterns_F^A$, list of access patterns to array A in the loop F

Communication Specifiers

Communication specifiers abstract calls to the runtime routines of gather and scatter types. By gather, we mean a routine which, before entering a data parallel loop, fetches the non–local elements referred to in the loop. By scatter, we mean a routine which, after a data parallel loop, updates the non–local elements modified by the loop.

```
PARAMETER :: N1 = ..., N2 = ..., K = ...
DOUBLE PRECISION, DIMENSION (N1), DISTRIBUTED (BLOCK) :: A
DOUBLE PRECISION, DIMENSION (N2), DYNAMIC :: B
INTEGER, DIMENSION (N1,2), DISTRIBUTED (BLOCK (K),:) :: X
INTEGER, DIMENSION (N2), DISTRIBUTED (BLOCK) :: MAP
   ...
READ (iu) MAP
DISTRIBUTE B :: (INDIRECT (MAP))
   ...
       DO step = 1, maxstep
F:         FORALL i = 1, N1 ON OWNER (A(i))
               A(i) = B(i) + B(X(i,1)) + B(X(i,2))
           END FORALL
       END DO
```

(a) Input Code Fragment

L: DD^X := Comp_DD (influencers of DD^X)
L: DD^B := Comp_DD (influencers of DD^B)
E: ESD_F := Comp_ESD (1, N1, 1, on_owner, DD^A, ii_j)
D: DAD^B := Comp_DAD ((ESD_F, DD^B, $< i,X(i,1),X(i,2) >$))
G: B   := Gather (DAD^B, B, DD^B))
F: ...FORALL loop ...

(b) After Insertion of the Specifiers

Fig. 4.56. Insertion of Descriptors

The communication specifiers for array A occurring in the **FORALL** loop F have the following structure:

$$
\begin{aligned}
&A := \text{Gather } (A, \text{DAD}_F^A, \text{DD}^A) \\
&A := \text{Scatter } (A, \text{DAD}_F^A, \text{DD}^A) \\
&A := \text{Scatter\_Add } (A, \text{DAD}_F^A, \text{DD}^A) \\
&A := \text{Scatter\_Mult } (A, \text{DAD}_F^A, \text{DD}^A)
\end{aligned}
$$

The meaning of the specifier influencers DAD_F^A and DD^A is clear from the previous text.

The use of the specifiers discussed above is illustrated in context of Fig. 4.56. The input code fragment (part (a)) is derived from the AVL FIRE kernel discussed in Section 4.3. Construction of data distribution descriptors is only outlined in this figure.

```
       PROCESSORS PROC(M)
       REAL, DIMENSION(10000), DISTRIBUTED (BLOCK) TO PROC ::      &
           A, B, C, D, E, F, G, H
       INTEGER, DIMENSION(10000) :: X, Y, Z
       . . .

L1:    FORALL i = 1, N ON OWNER (A(X(i)))
           A(X(i)) = B(Y(i))
       END FORALL

       DO j = 1, maxj
L2:        FORALL i = 1, N ON OWNER (C(X(i)))
               C(X(i)) = D(Z(i))
           END FORALL
       END DO

       IF (u > v) THEN
L3:        FORALL i = 1, N ON OWNER (E(X(i)))
               E(X(i)) = F(Y(i))
           END FORALL
       ELSE
           DO k = 1, maxk
L4:            FORALL i = 1, N ON OWNER (G(X(i)))
                   G(X(i)) = H(Y(i))
               END FORALL
           END DO
       END IF
```

Fig. 4.57. Example Code – Input to Intraprocedural Optimization

4.5.6 Intraprocedural Optimizations

In this subsection, we introduce a method for the elimination of the partial redundancy of the work distribution computation, communication precom-

putation, and communication by the optimal placement of their specifiers
and subsequent elimination of redundant specifiers.

The code fragment in Fig. 4.57 will be used for the illustration of the
power of this method. In the initial parallelization phase, the specifiers are
introduced into the IR; after this transformation, the executive part of our
example code has the form introduced in Fig. 4.58; the corresponding flow
graph is introduced in Fig. 4.59. In the communication specifiers of gather
type, we omit the data distribution descriptor operands.

```
        ESD_L1 = Comp_ESD (1, N, 1, on_owner_type, dd_A, <X(i)>)
        DAD_B = Comp_DAD (ESD_L1, dd_B, <Y(i)>)
        B = Gather (DAD_B, B)

L1:     FORALL i = 1, N ON OWNER (A(X(i)))
            A(X(i)) = B(Y(i))
        END FORALL

        DO j = 1, maxj
            ESD_L2 = Comp_ESD (1, N, 1, on_owner_type, dd_C, <X(i)>)
            DAD_D = Comp_DAD (ESD_L2, dd_D, <Z(i)>)
            D = Gather (DAD_D, D)

L2:         FORALL i = 1, N ON OWNER (C(X(i)))
                C(X(i)) = D(Z(i))
            END FORALL
        END DO

        IF (u > v) THEN
            ESD_L3 = Comp_ESD (1, N, 1, on owner type, dd_E, <X(i)>)
            DAD_F = Comp_DAD (dd_F, ESD_L3, dd_F, <Y(i)>)
            F = Gather (DAD_F, F)
L3:         FORALL i = 1, N ON OWNER (E(X(i)))
                E(X(i)) = F(Y(i))
            END FORALL
        ELSE
            DO k = 1, maxk
                ESD_L4 = Comp_ESD (1, N, 1, on_owner_type, dd_G, <X(i)>)
                DAD_H = Comp_DAD (ESD_L4, dd_H, <Y(i)>)
                H = Gather (DAD_H, H)
L4:             FORALL i = 1, N ON OWNER (G(X(i)))
                    G(X(i)) = H(Y(i))
                END FORALL
            END DO
        END IF
```

Fig. 4.58. Example Code – After Insertion of Specifiers in the Initial Parallelization
Phase

We can observe that in both Fig. 4.58 and 4.59, several ESD (DAD) descriptor objects are used for one and the same computational pattern. There is no attempt to reuse objects introduced for the same purpose earlier. For example, two different execution set descriptors, ESD_L1 and ESD_L4 are computed for loops L1 and L4 using the same computation pattern.

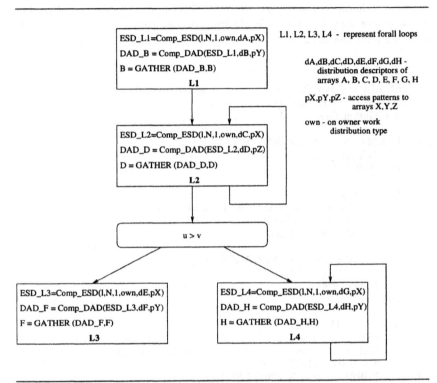

Fig. 4.59. Example Flow Graph – After Insertion of Specifiers

Because all these different objects are also *used* in subsequent candidate computations (i.e., are amongst their operands), this leads to an undesirable growth of the number of patterns, increasing complexity of the recognition of redundancies. Fortunately, this can be avoided by using one canonical ESD (DAD) object for the multitude of ESD (DAD) objects possibly used in the unoptimized program for one and the same ESD (DAD) computation pattern. This is feasible because there is no need to have more than one such object per pattern: The lifetimes of two ESD (DAD) objects assigned by the same ESD (DAD) computation pattern but under different execution contexts (and therefore carrying different values) will never overlap.

Or, in other words, since the functionality needed from an ESD (DAD) object is exactly the one provided by an expression temporary (see Section

4.5.2) – to carry a value from a definition to a use (or several uses) along invalidation–free paths only – it is, as with expression temporaries, sufficient to employ exactly one object per pattern.

That way, from the PRE point of view, each ESD (DAD) object can be perfectly regarded as an expression temporary for the candidate computation pattern it belongs to.

The situation described above inspires a method for a) identifying IR subtrees as ESD (DAD) – level expression patterns while b) normalizing the program to employ one single canonical target object per computation pattern, with the net effect that c) it is guaranteed that all uses of ESD/DAD objects will be correct under the new placement as well. This method is described in more detail in work [84] and basically works as follows:

- One canonical target object is determined for each pattern.
- Definitions of other targets in candidates that are identified with the same pattern are replaced by definitions of the canonical target object.
- All uses of such 'alternate' targets are replaced by references to the corresponding canonical target.

After applying this transformation we obtain the flow graph depicted in Fig. 4.60. This graph is used as input to PRE algorithms.

When using the classic PRE code motion techniques(e.g., [102, 192]) it is possible to obtain a *computationally optimal* solution, i.e. the solution in which the number of computations on each program path cannot be reduced anymore by means of the *safe* code motion[17]. Central idea to obtain this optimality result is to place computations as *early as possible* in a program, while maintaining safety. However, this strategy moves computations even if it is *unnecessary*, i.e. there is no runtime gain. Because computations whose placement we optimize usually generate large data structures that implement the descriptor objects this approach causes superfluous *memory pressure* which may be a serious problem in practice. When parallelizing large scientific applications these large temporary data structures may cause that the SPMD program will run out of memory. This is even more important on the machines (e.g., Parsytec PowerGC) which do not support virtual memory. Therefore, there is the need for a method that is as efficient as traditional PRE methods and avoids any unnecessary memory pressure.

Further, a code motion algorithm is needed that captures all *second order effects* between transformations. For example, hoisting the ESD computation from block 5 of Fig. 4.60 enables hoisting the DAD computation and consequently the GATHER communication from this block.

---

[17] Recall that code motion is safe when it doesn't introduce computations of new values on paths of the flow graph.

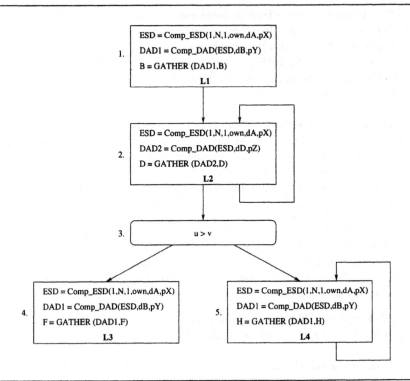

Fig. 4.60. Example Flow Graph – After Introduction of Canonical Descriptor Objects

The both issues mentioned are addressed by Knoop et al. in [161, 164], in the context of expression and assignment motion optimizations. In [161] they present a *lazy* computationally optimal code motion algorithm, which avoids any unnecessary memory pressure while maintaining computational optimality. This idea is subsumed in work [164], where a powerful algorithm for assignment motion is presented that for the first time captures all second order effects between assignment motion and expression motion transformations. In addition to covering second order effects, their algorithm captures arbitrary control flow structures and elegantly solves the usual problem of distinguishing between profitable code motion across loop structures and fatal code motion into loops. We use the power of the algorithm[18] introduced in [164] to placement and elimination of specifiers. The presentation style and part of the notation used in the rest of this section are strongly based on work [164].

---

[18] The original form of the algorithm is only applied to assignment statements whose *rhs* contains at most one operator symbol. Therefore, it is necessary to adapt this approach to our *"ESD–DAD–communication specifiers"* environment.

Preliminaries

The *program flow graph* for a program P is a quadruple $G = (N, E, \mathbf{s}, \mathbf{e})$ where N is the set of nodes of the flow graph, each node $n_i \in N$ corresponding to a basic block of P, E is the set of edges representing transfer of control between the basic blocks, and $\mathbf{s}$ is the unique entry and $\mathbf{e}$ exit node of the program P. A node n_i is a *predecessor* of node n_j, expressed by $pred(n_j)$, iff there exists an edge (n_i, n_j) in E. Similarly, a node n_j is a *successor* of node n_i, expressed by $succ(n_i)$, iff there exists an edge (n_i, n_j) in E.

Let there be two sets of nodes N_i and N_j of the program flow graph G, $\Theta(N_i, N_j)$ denotes the set of all paths from any node $n_i \in N_i$ to any node $n_j \in N_j$.

A *program point* w is the instant between the execution of an statement S_n and the execution of the statement S_{n+1}. The effect of statement S_n is said to be completely realized at point w.

Nodes of the program flow graph represent basic blocks of instructions. Instructions are assignment statements included in the original program and specifiers of the form $v := t$ and other statements, for example, empty statement, I/O statements, branching statements, and communication statements. We consider *variables* v to be from the set of all program variables, and *terms* t to be from the set of all program terms. An *assignment (expression) pattern* α (ϵ) is a string of the form $v := t$ (t). We assume that every expression pattern ϵ is associated with a unique temporary h_ϵ which is used for storing the value of ϵ in order to eliminate partially redundant occurrences of ϵ. Given a program G, $\mathcal{EP}$ denotes the set of all expression patterns occurring in G, and $\mathcal{AP}$ the set of all assignment patterns which is enriched by the set of all assignment patterns of the form $h_\epsilon := \epsilon$ and $v := h_\epsilon$ for all $\epsilon \in \mathcal{EP}$ occurring on the left hand side of an assignment with left hand side variable v.

The Algorithm

The algorithm consists of the following three steps:

1. *Initialization:* Introducing temporaries
2. *Assignment Motion:* Eliminating partially redundant assignments
3. *Final Flush:* Eliminating unnecessary initializations of temporaries

The next parts describe the above steps in more detail and introduce the corresponding data flow equations. In the following, the term *assignment* will mean either a standard assignment or an update of the variable caused by the communication.

The Initialization Phase

The initialization phase – *Program transformation 1* – replaces every statement $x := t$ by the assignment sequence $h_t := t; x := h_t$, where h_t is the unique temporary associated with term t.

The Assignment Motion Phase

The assignment motion phase eliminates partially redundant assignments in the program resulting from the initialization phase. This is achieved in two steps: (a) eliminating redundant occurrences of assignments, and (b) moving the assignments as far as possible in the opposite direction of the control flow to their 'earliest' safe execution points. These two steps are applied until the program stabilizes.

Redundant Assignment Elimination. The elimination of redundant assignments is based on a forwards directed bit–vector data flow analysis introduced in Fig. 4.61 (a). Here

$$\text{N--REDUNDANT}_k(\alpha) \quad (\text{ or } \text{ X--REDUNDANT}_k(\alpha))$$

means that assignment pattern α is redundant at the entry (or exit) of assignment k.

After having computed the greatest solution of the equation system it is possible to perform the elimination step – *Program transformation 2* – which processes every basic block by successively eliminating all assignments that are redundant immediately before them, i.e., redundant at their entry.

Assignment Hoisting. Fig. 4.61 (b) presents the hoistability analysis in a bit–vector format, where each bit corresponds to an assignment pattern α, occurring in the program. Here, N--HOISTABLE_n and X--HOISTABLE_n intuitively mean that some hoisting candidates of α can be moved to the entry or exit of the basic block n, respectively, where a *hoisting candidate* is an occurrence of an assignment $x := t$ inside a basic block which is not blocked, i.e., neither preceded by a modification of an operand of t nor by a modification or a usage of x.

The greatest solution of the equation system displayed in Fig. 4.61 (b) characterizes the program points, where instances of the assignment pattern α must be inserted, by means of the insertion predicates N--INSERT_n and X--INSERT_n. The insertion step – *Program transformation 3* – processes every basic block by successively inserting instances of every assignment pattern α at the entry (or exit) of n if $\text{N--INSERT}_n(\alpha)$ (or $\text{X--INSERT}_n(\alpha)$) is satisfied, and simultaneously removes all hoisting candidates.

Local Predicates: (Let $\alpha \equiv v := t \in \mathcal{AP}$ such that v is not an operand of t.
Let s denote the first instruction of s).

EXECUTED$_i(\alpha)$:   Instruction i is an assignment of the pattern α.
LOC-BLOCKED$_n(\alpha)$:   Instruction i is transparent for α, i.e. neither v nor
any operand of t is modified by i.

Redundancy Analysis:

$$N - \text{REDUNDANT}_i \;\; =_{df} \begin{cases} false & \text{if } i = s \\[2mm] \prod_{i \in pred(i)} X - \text{REDUNDANT}_i & \text{otherwise} \end{cases}$$

$$\text{X-REDUNDANT}_i \;=_{df}\; \text{EXECUTED}_i + \text{ASS-TRANSP}_i * \text{N-REDUNDANT}_i$$

(a) Redundant Assignment Analysis

Local Predicates: (Let $\alpha \equiv v := t \in \mathcal{AP}$)

LOC-HOISTABLE$_n(\alpha)$:   There is a hoisting candidate of α in n.
LOC-BLOCKED$_n(\alpha)$:   The hoisting of α is blocked by some
instructions of n.

Hoistability Analysis:

$$\text{N-HOISTABLE}_n \;=_{df}\; \text{LOC-HOISTABLE}_n + \text{X-HOISTABLE}_n * \overline{\text{LOC} - \text{BLOCKED}_n}$$

$$X - \text{HOISTABLE}_i \;\;=_{df} \begin{cases} false & \text{if } n \text{ is the} \\ & \text{entry block} \\[2mm] \prod_{m \in succ(n)} N - \text{HOISTABLE}_m & \text{otherwise} \end{cases}$$

Insertion Points:[a]

$$\text{N-INSERT}_n \;=_{df}\; \text{N-HOISTABLE}_n^* * \sum_{m \in pred(n)} \overline{X - \text{HOISTABLE}_m^*}$$
$$\text{X-INSERT}_n \;=_{df}\; \text{X-HOISTABLE}_n^* * \text{LOC-BLOCKED}_n$$

[a] N-HOISTABLE* and X-HOISTABLE* denote the greatest solution of the equations
for hoistability.

(b) Hoistability Analysis and Insertion Points

Fig. 4.61. Assignment Motion Phase

Local Predicates: (Let h_ϵ be the temporary for some $\epsilon \in \mathcal{EP}$.
Let s denote the first instruction of s.)

$\text{IS–INST}_i(h_\epsilon)$:   Instruction i is an instance of $h_\epsilon := \epsilon$.
$\text{USED}_i(h_\epsilon)$:   Instruction i uses h_ϵ.
$\text{BLOCKED}_i(h_\epsilon)$: Instruction i blocks $h_\epsilon := \epsilon$.

Delayability Analysis:

$$N-\text{DELAYABLE}_i \quad =_{df} \quad \begin{cases} \textit{false} & \text{if } i = s \\ \prod_{i \in pred(i)} X-\text{DELAYABLE}_i & \text{otherwise} \end{cases}$$

$$X\text{–DELAYABLE}_i =_{df} \text{IS–INST}_i + N\text{–DELAYABLE}_i * \overline{\text{USED}_i} * \overline{\text{BLOCKED}_i}$$

Usability Analysis:

$$N\text{–USABLE}_i =_{df} \text{USED}_i + \overline{\text{IS} - \text{INST}_i} * X\text{–USABLE}_i$$
$$X\text{–USABLE}_i =_{df} \sum_{i \in succ(i)} N - \text{USABLE}_i$$

Computing Latestness:

$$N\text{–LATEST}_i =_{df} N\text{–DELAYABLE}_i^* * (\text{USED}_i + \text{BLOCKED}_i)$$
$$X\text{–LATEST}_i =_{df} X\text{–DELAYABLE}_i^* * \sum_{i \in succ(i)} \overline{N - \text{DELAYABLE}_i^*}$$

Initialization Points: Insert instance of α

$$N\text{–INIT}_i =_{df} N\text{–LATEST}_i * N\text{–USABLE}_i^*$$
$$X\text{–INIT}_i =_{df} X\text{–LATEST}_i$$

Reconstruction Points: Reconstruct original usage of t instead of h

$$\text{RECONSTRUCT}_i =_{df} \text{USED}_i * N\text{–LATEST}_i * \overline{X - \text{USABLE}_i^*}$$

Eliminating Unnecessary Assignments to Temporaries

Fig. 4.62. Final Flush Phase

Remark: All assignment patterns that must be inserted at a particular program point are independent and can therefore be placed in an arbitrary order.

The Final Flush Phase

The final flush phase moves the occurrences of all assignment patterns of the form $h_\epsilon := \epsilon$ to their 'latest' safe execution points, and eliminates all occurrences whose left hand side is used at most once immediately after their occurrence. The procedure is based on two uni–directional bitvector data flow analyses computing *delayable* program points, where an initialization is *usable*. The data flow equations for the final flush phase are introduced in Fig. 4.62.

The flow graph of the example code after optimizations is outlined in Fig. 4.63.

Remark: Communication of scatter type is placed using sinking transformations as they are described in [162].

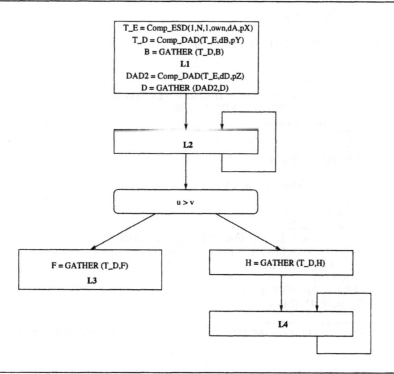

Fig. 4.63. Example Flow Graph – The Effect of the Optimization

4.5.7 Interprocedural Optimizations

So far, we addressed the problem of performing partial redundancy elimination intraprocedurally, by a separate and independent investigation of the procedures of a program. The purpose of the interprocedural method is to increase the effect of optimizations by determining the relevant effects of procedure calls and allowing code motion between procedures.

```
PROGRAM MAIN

    PROCESSORS R(4)
    REAL, DISTRIBUTED (BLOCK) TO R :: A1(100), A2(100), B(100)
    INTEGER X(1000), Y(1000)
    COMMON N
    ...
    sseq1
    DO k = 1, MAX
        CALL P (A1,B,X,Y)
        CALL Q (A2,B,X,Y)
    END DO
    sseq2
END PROGRAM MAIN

SUBROUTINE P(FA1, FB, FX, FY)
    ...
    sseq3
    DO it = 1, N
L1:         FORALL j = 2, 1000 DISTRIBUTE ( BLOCK )
                FA1(FX(j)) = FB(FY(j-1))
            END FORALL
    END DO
    FA1 = Func1(FA1)
END SUBROUTINE P

SUBROUTINE Q(FA2, FB, FX, FY)
    ...
    sseq4
    DO it = 1, N
L2:         FORALL k = 2, 1000 DISTRIBUTE ( BLOCK )
                FA2(FX(k)) = FB(FY(k-1))
            END FORALL
    END DO
    FA2 = Func2(FA2)
END SUBROUTINE Q
```

Fig. 4.64. Program Involving three Procedures – After Insertion of Specifiers

A possible solution is in–line expansion, a technique which textually replaces procedure calls by the appropriately modified procedure bodies. After this transformation, standard intraprocedural techniques can be applied to the resulting program. The drawback of this method is the 'potentially exponential' growth in program size which makes it generally infeasible for all but small programs [274].

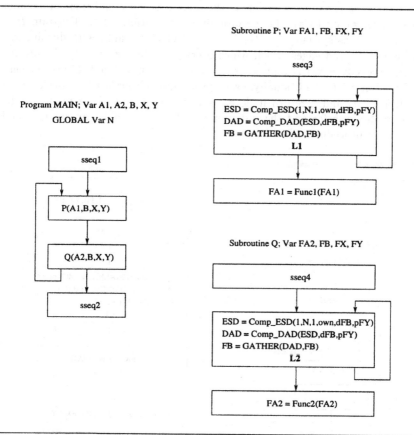

Fig. 4.65. Flow Graphs for Procedures MAIN, P, and Q

There is a great variety of papers dealing with specific problems of interprocedural DFA. The approach presented by Knoop et al. [163, 160] is the first one that solves the task of eliminating interprocedural partial redundancies optimally. This approach is general: it applies to a broad range of practically relevant problems, and it is precise even in the presence of recursive procedures.

In the next paragraphs, we outline the application of the Knoop's approach to the interprocedural optimizations of irregular codes.

Flow Graph System

In the interprocedural setting program Π is considered as a system $(\pi_0, \pi_1, ..., \pi_k)$ of (mutually recursive) procedure definitions, where every procedure π of Π has a list of formal value and reference parameters, and a list of local variables. π_0 is assumed to denote the main procedure and therefore cannot be called. π_1 up to π_k are the procedure declarations of Π. Program Π is represented as system $S = (G_0, G_1, ..., G_k)$ of flow graphs with disjoint sets of nodes N_i and edges E_i, in which every procedure π of Π (including the main procedure π_0) is represented as a directed flow graph G in the sense of Section 4.5.6. Additionally, every procedure call node n is replaced by a call node n_c and a return node n_r, which are introduced for modeling the parameter transfer and the initialization of local variables on entering and leaving a procedure called by n.

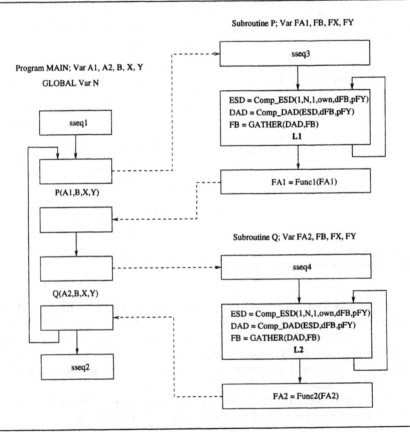

Fig. 4.66. Flow Graph System

Optimization

In Fig. 4.64, we show one example program which involves irregular accesses. The system $S = (G_0, G_1, G_2)$ of three flow graphs that corresponds to our example program after insertion of specifiers is shown in Fig. 4.65 and the corresponding flow graph system is depicted in Fig. 4.66.

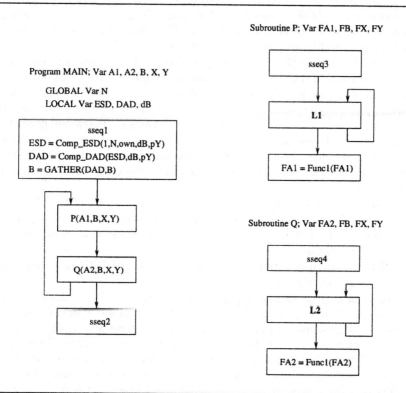

Fig. 4.67. After Interprocedural Optimization

The generic algorithm for computing interprocedural data flow solutions introduced in [163] is decomposed into a *preprocess*, computing the effects of procedure calls, and a *main process*, which subsequently computes the desired data flow solution. In [160], the specifications of the interprocedural analyses for assignment sinking and assignment hoistings, the central analyses for partial dead code elimination, partially redundant assignment elimination, are presented. Application of these analyses to interprocedural placement of specifiers introduced in Section 4.5.6 is presented in work [58]. In Fig.

4.67, we show the flow graph system after application of this interprocedural optimization to the flow graph system depicted in Fig. 4.66.

BIBLIOGRAPHICAL NOTES

An iterative intraprocedural data flow framework for generating communication statements in the presence of indirection arrays has been developed by Hanxleden et al. [259]. It uses a loop flow graph designed to ease the hoisting of communication out of innermost loops. However, it handles nested loops and jumps out of loops only to a limited degree.

Later, Hanxleden [258] has developed Give–N–Take, a new intraprocedural data placement framework. This framework extends PRE in several ways, including a notion of early and lazy problems, which is used for performing earliest possible placement of receive operations. Allowing such asynchronous communication can reduce communication latencies. Hanxleden [256] describes the implementation of Give–N–Take method in the prototype of the Fortran D compiler. Inspectors are hoisted out of loops only as far as the communications are moved out. Optimization of the work distributor is not considered.

Nadeljkovic and Kennedy [157] show that combination of the Give–N–Take method and data dependence analysis, traditionally used to optimize regular communication, allows to perform more extensive optimizations of regular computations.

PRE was also used intraprocedurally by Gupta et al. [137] for performing communication optimizations.

There is a great variety of papers dealing with specific problems of interprocedural DFA. Most of these methods do not properly deal with local variables of recursive procedures. For example, Morel and Renvoise [193] proposed an algorithm for the interprocedural elimination of partial redundancies. Their solution is heuristic in nature, and their work is restricted to the programs whose call graph is acyclic. They also do not consider the possibility that the procedure having a candidate for placement may be invoked at multiple call sites with different sets of parameters and do not maintain accuracy of solutions when procedures are invoked at multiple call sites.

The first efforts on interprocedural analysis for distributed memory compilation are by Gerndt [118] and Hall et al. [140]. They have concentrated on flow–insensitive analysis[19] for regular applications, including management of buffer space and propagation of data distribution and data alignment information across procedure boundaries. Framework for Interprocedural Analysis and Transformation (FIAT) [67] has recently been proposed as a general environment for interprocedural analysis.

---

[19] Flow–insensitive analysis does not take into account the set of all control paths inside the procedure.

A new method for schedule placement at the interprocedural level is presented in [3, 4]. It is based on application of the interprocedural partial redundancy elimination. A directed multigraph referred as Full Program Representation is constructed and used for passing of data flow information at the inter- and intra-procedural level. However, they consider communication operations for interprocedural placement only if this operation is not enclosed by any conditional or loop. Furthermore, they do not consider any second order effects.

5. Compiling Parallel I/O Operations

The language Vienna Fortran 90 includes both standard Fortran 90 I/O operations and the explicit parallel I/O operations specified by the annotation introduced in Section 3.3.

The implementation of all I/O operations comprises both runtime and compile time elements. Using a set of interface procedures provided by the I/O runtime system VIPIOS (VIenna Parallel Input-Output System) we can compile all the Vienna Fortran 90 I/O operations for distributed-memory systems.

In this chapter, we concentrate on the compile time features of the implementation of the explicit parallel I/O operations; handling standard Fortran 90 operations within the VFCS is described in [54]. In Chapter 7, we deal with the runtime support issues.

Processing of parallel I/O operations can be conceptually done in two ways:

1. *Direct I/O compilation*

   This approach extracts parameters about data distributions and file access patterns from the Vienna Fortran programs and passes them in a normalized form to the VIPIOS runtime primitives without performing any sophisticated program analysis and optimizations.

2. *Optimizing I/O compilation*

   Although the resulting program produced by the direct compilation method is supported by the parallel I/O runtime system, it may still be I/O inefficient. Aggressive program optimizations techniques are to be applied in order to improve its I/O efficiency. There are several possibilities how to increase the execution speed of the application program; for example, by executing I/O operations in parallel with computation and other I/O operations.

   The program optimization is based on program analyses and appropriate program transformations. Like in the case of communication optimizations, it is suitable to support analysis and transformations by high level information structures, called I/O communication descriptors. In the initial parallelization phase, an I/O communication descriptor is constructed and inserted into the Internal Representation (IR) for each parallel I/O communication. I/O communication descriptors are abstractions of the actual I/O operations and are expanded into calls to appropriate VIPIOS primitives in the code generation phase. I/O communication descriptors contain the information provided by the user, information gathered by the program analysis, and may be extended by the performance information collected and stored in the IR by the execution monitor (performance analysis) during the program run.

Sections 5.1 and 5.2 deal with the above compilation approaches in greater detail.

5.1 Direct I/O Compilation

This section describes the basic compilation methodology used in the VFCS when translating I/O operations on parallel files. The compiler translates these operations using the following steps:

1. Analyze the I/O communication parameters.
2. If the statement includes distributed arrays, determine the distribution of these arrays using the interprocedural data distribution analysis that is provided within the VFCS. Initialize the corresponding data distribution descriptors.
3. Insert calls to the VIPIOS runtime library and pass them the arguments required.

Below we describe the translation of the basic Vienna Fortran concurrent I/O operations. In the translation examples introduced, we do not present the complete VIPIOS interface; we omit, for example, return arguments, that return a success code or a failure code. Consequently, we do not include any correctness checks in the code generated. Specification of the complete VIPIOS interface can be found in [59].

We use *keyword arguments* in procedure calls. This makes possible to put the arguments in any order and even to call the procedures with optional arguments.

5.1.1 OPEN Statement

Specifications occurring in the statement are evaluated and stored in variables that are passed to the function VIPIOS_open which returns a file descriptor. This is shown in the following example.

Example: Translation of the OPEN statement that opens a new file

---

! Original code

```
DFUNCTION PAT4(M,N)
    TARGET_PROCESSORS P2D(M,M)
    TARGET_ARRAY, ELM_TYPE (REAL) :: A (N,N)
    ! regular distribution type
    A DISTRIBUTED (BLOCK, BLOCK) TO P2D
END DFUNCTION PAT4
```

```
OPEN (13, FILE ='/usr/exa6', FILETYPE ='PAR', STATUS ='NEW', &
             TARGET_DIST = PAT4(8,800))
```

! Transformed form (generated by the VFCS automatically)

```
PARAMETER :: Max_Numb_of_Units = ...
TYPE (Distr_Descriptor) :: dd
TYPE (File_Descriptor) :: fd
TYPE (File_Descriptor), DIMENSION (Max_Numb_of_Units) :: FdArray
          . . .
... Initialization of descriptor dd ...
CALL   VIPIOS_open(name='/usr/exa6',status='NEW', dist_descr=dd, &
                        file_desc=fd)
FdArray(13) = fd
```

Data distribution descriptor denoted by *dd*, is initialized by the data that is derived evaluating the TARGET_DIST specification. File descriptor *fd* stores all information about the file in the compact form. This information is needed in the subsequent file operations. File descriptors are stored in one dimensional array FdArray using the file unit number as an index. FdArray is initialized with dummy values during the SPMD program prolog phase, and it can be available to each program unit through the COMMON area or MODULE mechanisms, for example.

Opening an Old File. If the file that is being open has status 'OLD', and a TARGET_DIST specification is introduced, the opening may be associated with a file reorganization if the old TARGET_DIST specification does not equal the new one.

Example: Translation of the OPEN statement that opens an old file

---

! *Original code*

```
OPEN (13, FILE ='/usr/exa6', FILETYPE ='PAR', STATUS ='OLD',&
                    TARGET_DIST = PAT4(4,800))
```

! *Transformed form*

```
PARAMETER :: Max_Numb_of_Units = ...
TYPE (Distr_Descriptor) :: dd
TYPE (File_Descriptor) :: fd_old, fd_new
TYPE (File_Descriptor), DIMENSION (Max_Numb_of_Units) :: FdArray
        . . .
... Initialization of descriptor dd ...
CALL   VIPIOS_open(name='/usr/exa6',status='OLD',file_desc=fd_old)
IF (ISCFILE(fd_old) .AND. IODIST(fd_old) .NEQ. dd) THEN
     CALL   VIPIOS_reorganize(file_descr_old=fd_old, &
                            file_descr_new=fd_new, dist_new=dd)
     FdArray(13) = fd_new
ELSE
     FdArray(13) = fd_old
END IF
```

Intrinsic function $IODIST$ takes a file descriptor as an argument and returns a data distribution descriptor which is a component of the file descriptor. $ISCFILE$ is a logical function that checks whether the file descriptor passed to it as an argument describes a parallel file.

5.1.2 WRITE Statement

The WRITE statement is translated to a call of function VIPIOS_write that writes synchronously the distributed array occurring in the statement to the open VIPIOS file.

Example: Translation of the WRITE statement

! Original code

    ... processor specification ...
    ... declaration of A, data distribution specification for A
    ... open of f ...

WRITE (f) A

! Transformed form

TYPE (Distr_Descriptor) :: dd_source
TYPE (File_Descriptor), DIMENSION (Max_Numb_of_Units) :: FdArray

    . . .
... Initialization of dd_source according to the distribution of A ...
IF (*ISCFILE*(FdArray(f))) THEN
    CALL VIPIOS_write (file_descr=FdArray(f),data_address = A, &
                                      dist_source=dd_source)
ELSE IF (*ISDUMMY*(FdArray(f))) THEN
    PRINT *, 'No open File is attached to unit', f
ELSE
    PRINT *, 'Attempt to apply parallel operation to a standard file'
END IF

ISDUMMY is a logical function that checks whether the argument represents a dummy file descriptor.

If a TARGET_DIST specification is introduced in the WRITE statement the additional parameter dist_target is passed to VIPIOS_write, as it is shown below.

! Original code

    ... processor specification ...
    ... declaration of A, data distribution specification for A
    ... open of f ...

WRITE (f, TARGET_DIST = PAT4(16,800)) A

!Transformed form

TYPE (Distr_Descriptor) :: dd_source, dd_target
TYPE (File_Descriptor), DIMENSION (Max_Numb_of_Units) :: FdArray
    . . .
... Initialization of dd_source according to the distribution of A ...
... Initialization of dd_target according to the TARGET_DIST
    specification ...
CALL   VIPIOS_write (file_descr=FdArray(f),data_address = A,   &
                            dist_source=dd_source,dist_target=dd_target)

▲

5.1.3 READ Statement

The READ statement is translated to a call of function VIPIOS_read that synchronously reads data from the open VIPIOS file into the distributed array occurring in the statement.

Example: Translation of the READ statement

▼

! Original code

READ (f) A

! Transformed form

TYPE (Distr_Descriptor) :: dd
TYPE (File_Descriptor), DIMENSION (Max_Numb_of_Units) :: FdArray
... Initialization of dd according to the distribution of A ...
CALL   VIPIOS_read (file_descr=FdArray(f),data_address = A, &
                            dist_target=dd)

▲

5.2 Optimizing I/O Compilation

The aim of the optimizing compilation, discussed in this section is to improve the I/O performance of the parallel program through compile time program manipulations and by deriving additional information that is passed to the runtime system as the basis for runtime optimizations. Optimizations are based on the results of program analysis which are provided by the Analysis Subsystem of the VFCS.

The goal of I/O optimizations is to derive an efficient parallel program with

- *Low I/O overhead*
  The compiler derives hints for data organization in the files and inserts the appropriate data reorganization statements.
- *High amount of computation that can be performed concurrently with I/O*
  These optimizations restructure the computation and I/O to increase the amount of useful computation that may be performed in parallel with I/O. The VIPOS calls offer the compiler the choice of synchronous or asynchronous I/O. To overlap I/O with computation, the asynchronous mode is used.
- *I/O performed concurrently with computation and other I/O*

Program analysis that supports I/O optimizations comprises three conceptually distinct issues that are described below.

1. *Data Flow Analysis*
   It examines the flow of scalar values through a program. Accesses of array elements are simply modeled as accesses to the entire array. This is adequate for optimizing I/O operating on whole arrays. Reading from or writing to any logical file is treated like reading from or writing to a single variable.
2. *Reaching Distribution Analysis*
   In order to be able to generate efficient code for parallel I/O operations, a compiler needs information about the distribution or as much information as possible about the set of all distributions that the array may be associated with (the distribution range) that reach a given point of reference to the array. The solution of the reaching distributions problem has been implemented in VFCS at the interprocedural level.
3. *Cost Estimation*
   The data flow and reaching distribution analyses are insufficient to automatically derive effective hints for the data layout and select efficient program restructuring for arbitrary programs. The compiler can only empirically apply and superficially evaluate different program transformations. This problem has been significantly relieved by an appearance of performance prediction tools. Their goal is to provide performance feedback for all compiler decisions in order to guide the restructuring process through the search tree and detect the most efficient program version. Such a tool is able to predict the execution time of a program version or code part. Another important performance prediction issue is the determination of system specific I/O cost parameters.

Processing parallel I/O operations which includes compile time optimizations conceptually consists of four phases:

- Insertion of I/O communication descriptors in the initial parallelization phase
- Program analysis and enrichment of I/O communication descriptors with some results of the analysis
- Optimizing transformations
- Expansion of I/O communication descriptors into VIPIOS calls

Next, we deal with the I/O communication descriptors and a number of I/O optimizations.

5.2.1 I/O Communication Descriptors

In order to allow manipulation of I/O data movement at a rather early stage of parallelization, it is necessary to explicitly describe possible data movements by means of high level I/O communication descriptors. A high level I/O communication descriptor[1] contains all information that is needed to determine which elements have to be transferred between processors' local memories and disks and how the transfer is organized. An important aspect of the high level communication descriptor is the abstraction from the details of the actual management of temporary storage and I/O transfer.

I/O communication descriptors must be flexible enough to allow optimization operations of them; for example, recognizing and removing redundant descriptors, movement of descriptors (i.e. extraction from loops, across procedure boundaries), fusion of descriptors, and splitting into reading/writing and waiting components in case of asynchronous operations.

The high level I/O communication descriptor PIO_COMM has the following structure:

> PIO_COMM (iotype, f, A, dd_A, [, section_desc_A] [, hints] [, cost])

with

- *iotype*, type of I/O communication (e.g., *read*)
- *f*, file unit
- *A*, array whose elements are to be transferred
- *dd_A*, distribution descriptor of *A*
- *section_desc_A*, description of the section of *A* to be transferred
- *hints*, the list of hints provided by the user and/or derived by the compiler
- *cost*, system specific cost parameters

The use of descriptors will be illustrated in Subsection 5.2.2.

---

[1] Recall that read and write I/O operations can be considered as receive and send communication operations.

5.2.2 Automatic Determination of I/O Distribution Hints

If the automatic I/O mode is used, the programmer provides no I/O directives and hints; all decisions concerning optimizations are taken by the programming environment.

Consider the code fragment in Fig. 5.1. In this example, arrays having various distributions are transferred to/from the file *'matrix.dat'*. First, array A having BLOCK distribution is written to this file, then the file data is *ten* times read onto array C having CYCLIC(M) distribution, and finally this data is read *once* onto array B having CYCLIC distribution; the number of the respective READ operations, *ten* and *one*, is called *frequency* below. This information can be extracted from the code at compile time and stored as a hint in the I/O communication descriptor that replaces the WRITE statement *S1* in the program IR. This descriptor has then the following concrete form:

$$\boxed{\text{PIO\_COMM } (write, \text{ f, A, dd\_A, } <(10,\text{dd\_C}), (1,\text{dd\_B})>)}$$

The list of hints, each of them having the form *(frequency, data_distribution_descriptor)*, is marked by " $<$ " and " $>$ ".

```
INTEGER, PARAMETER :: M = ..., N = ..., f = ...
PROCESSORS P1D(M)
REAL, DISTRIBUTED ( BLOCK ) :: A(N)
REAL, DISTRIBUTED ( CYCLIC ) :: B(N)
OPEN ( UNIT =f, FILE ='/usr/matrix.dat', FILETYPE ='PAR', STATUS ='NEW')
        ...
S1: WRITE (f) A
DO  i = 1, 5
       CALL SUB(f, N, ...)
END DO
        ...
S2: READ (f) B
        ...
SUBROUTINE SUB(f, N, ...)
     INTEGER f
     REAL, DISTRIBUTED ( CYCLIC (M)) :: C(N)
           ...
     DO  j = 1, 2
           ...
           READ (f) C
           ...
     END DO
           ...
END SUBROUTINE SUB
```

Fig. 5.1. Arrays Having Various Distributions are Transferred to/from the File

5.2.3 Overlapping I/O with Computation

The VIPIOS calls offer the compiler the choice of synchronous or asynchronous I/O. To overlap I/O with computation, the asynchronous mode is used. The transformation of the READ statement based on the utilization of the asynchronous mode is shown in the example below. In this example, we do not introduce the intermediate program form that would include the I/O communication descriptor.

! Original code

```
READ (f) A
Statement_Sequence1       ! A does not occur in this part
Statement_Sequence2       ! A is used in this part
```

! Transformed form

```
INTEGER :: iready, result
TYPE (Distr_Descriptor) :: dd
TYPE (File_Descriptor), DIMENSION (Max_Numb_of_Units) :: FdArray
    . . .

... Initialization of dd according to the distribution of A ...
CALL  iVIPIOS_read (file_descr=FdArray(f),data_address = A,   &
                         dist_target=dd, event=iready)
Statement_Sequence1
CALL  VIPIOS_wait (iready)
Statement_Sequence2
```

In the above example, computation in *Statement_Sequence1* hides the read operation latency.

5.2.4 Running I/O Concurrently with Computation and Other I/O

Restructuring I/O statements in a program to run in parallel with other computations, and/or with other I/O statements, is a relatively new concept, one that compilers parallelizing programs for massively parallel systems have not considered by now. We will call this kind of parallelism as the *I/O-computation-I/O parallelism* in the following.

The program analysis is capable of providing information whether the I/O-computation-I/O parallelism is save (due to the data dependence analysis) and useful (due to the performance analysis). If both preconditions are fulfilled, the compiler allows I/O to run in parallel with other computations or other I/O statements.

In the basic compilation phase, all I/O statements are treated as synchronization points and it was ensured that they do not run in parallel with other statements. In Section 5.2.3, we have shown that it can be the advantage to allow some I/O statements to run in parallel with other computations using the asynchronous VIPIOS operations.

Sridharan et al. [241] investigate the feasibility and the benefits of allowing I/O-computation-I/O parallelism at the SPMD program level. Detecting of parallelism in this context is based on the dependence analysis. Reading from or writing to any file is treated like reading from or writing to a single variable. This variable is called a logical file. By default, all I/O statements are assigned to one single logical file. This dependence caused by this logical file forces different I/O statements to be serialized, but does not prevent I/O statements from running in parallel with other computations, if possible.

The following examples outlines the preconditions for running two WRITE statements in parallel.

```
OPEN ( UNIT = u1, ...); OPEN ( UNIT = u2, ...)
WRITE (u1) A; WRITE (u2) B
```

In the above code segment the two WRITE statements can be run in parallel either if they write to different logical files, or if the order of the output really does not matter (the file was opened with the specifier ORD = 'NO'). This control structure can be easily implemented by any lightweight thread package (for instance [138])).

Next, we study a simple multiple-matrix multiplication program with the following characteristics[2]:

1. Initial values for two matrices A and B are read from two different files $f1$ and $f2$, respectively.
2. Matrix C is initialized to zero.
3. The matrix product $C = A * B$ is computed.
4. Matrix C is written to file $f3$.
5. Matrix D is read from file $f2$.
6. Matrix E is initialized to zero.
7. The matrix product $E = C * D$ is computed.
8. Matrix E is is written to file $f3$.

In this simple program, I/O statements can run in parallel with other I/O statements, and with other computations. Depending on the capabilities of

---

[2] This is a modification of the example introduced in [241].

the target system, the compiler can optimize the scheduling. For example, if there is only a single I/O processor, the compiler can reduce the degree of I/O parallelism by preventing an I/O statement from running in parallel with other I/O statements.

A possible parallel control structure of the program is depicted in Fig. 5.2 where, *fork*, *join* and *barrier* denote creation, termination and synchronization of processes, respectively.

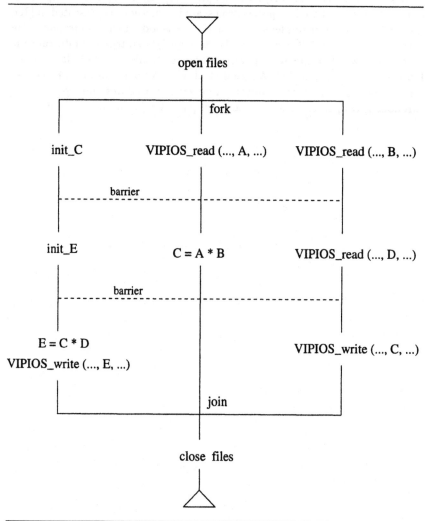

Fig. 5.2. Running I/O Concurrently with Computation and Other I/O

BIBLIOGRAPHICAL NOTES

So far, there has been little research in compiling I/O operations for distributed-memory systems. Bordawekar and Choudhary present language support (as extensions to HPF), runtime primitives, and initial compiler implementation in the Fortran90D/HPF compiler [41]. A technique for compilation of I/O statements based on polyhedron scanning methods is presented in [82]; however, it has been implemented for the host-node architecture only. The PASSION compiler [78] translates the Fortran I/O statements into runtime procedures which implement parallel accesses to files. HPF-1 data distributions are only supported. Agrawal et al. [7] developed an interprocedural framework, which performs analysis on program procedures and replaces synchronous I/O operations with a balanced pair of asynchronous operations.

6. Compiling Out-of-Core Programs

The motivating idea for the development presented in this chapter is that a compiler can automatically transform an appropriate in-core (IC) algorithm to operate on out-of-core (OOC) data through splitting array element sets into portions, partitioning (tiling) the computation and I/O insertion. To do this, the compiler will use the data distribution and OOC annotation introduced in 3.4 and static program analysis.

In order to translate OOC programs, in addition to the steps used for in-core compilation, the compiler also has to schedule explicit I/O accesses to disks.

Consider the simple Vienna Fortran program[1] shown in Fig. 6.1a. The running instance of the corresponding in-core SPMD program would have on processor $PROC(1)$, for example, the following form:

$$\text{DO } I = 1, 8$$
$$A(I) = B(I)$$
$$\text{END DO}$$

Because arrays A and B are aligned, no communication statements have to be inserted into the SPMD form. We assume that the available memory per each processor is two (8 on all processors altogether). Thus the overall computation will require four phases. The resultant stripmined code is shown in Fig. 6.1b. The outermost j loop forms the *stripmining* loop. Each iteration of the stripmining loop reads the tile of array B and stores the tile of array A.

The rest of this chapter is organized as follows. Section 6.1 describes models for OOC computations. The main features of the OOC compilation strategy are outlined in Section 6.2. Section 6.3 characterizes OOC codes considered in this chapter. Section 6.4 deals with compiling OOC regular codes and Section 6.5 with compiling OOC irregular codes.

---

[1] A similar explanatory example is used in [40].

PROCESSORS PROC(4)
! Assume that the available memory per each processor is two
REAL, DISTRIBUTED (BLOCK), OUT_OF_CORE, IN_MEM (8) :: A(32), B(32)

A(1:32) = B(1:32)

(a)

! The compiler generates a parameterized OOC SPMD program; its running
! intance on processor PROC(1), for example, will have the following form
! (the overall computation will require four phases):

! stripmining loop
DO   J = 1, 4
        ... routine for reading tile of B ...
        ! computational loop
        DO   I = 1, 2
            A((J-1*2+I) = B((J-1)*2+I)
        END DO
        ... routine for writing tile of A ...
END DO

(b)

Fig. 6.1. Out-of-core Compilation Approach.

6.1 Models for Out-of-Core Computations

Vienna Fortran out-of-core programs are restructured by the VFCS into SPMD out-of-core programs which are executed on the target architecture. To support the restructuring process, it is necessary to present to the compiler a suitable abstraction of the program execution on the existing architectures. In this section, we define models on which the execution as well as the compilation of out-of-core Vienna Fortran programs is based.

As we introduced in Section 2.1, I/O subsystems of different massively parallel systems are often very architecturally different. There are some systems with a disk or disk array on each compute node (e.g. Meiko CS-2 and IBM SP-2), as illustrated by Fig. 6.2(b), or compute nodes are connected to a high bandwidth disk system through the communication network[2] (e.g. Intel iPSC/2, iPSC/860, PARAGON and NCUBE systems), as shown in Fig. 6.2(a).

---

[2] Or it may have a separate interconnection network (Parsytec Systems, Intel Touch Stone Delta System).

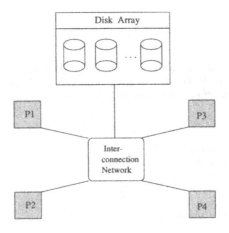

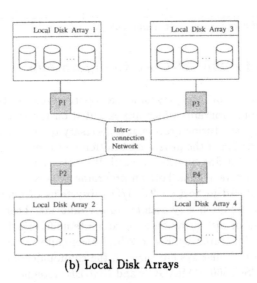

(b) Local Disk Arrays

Fig. 6.2. Shared vs. Distributed Disk Resources

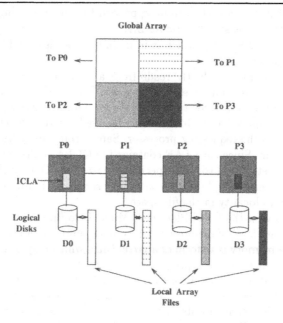

Fig. 6.3. Local Placement Model.

The models for OOC computations that are described in Subsections 6.1.1, 6.1.2, and 6.1.3 have been borrowed from Choudhary and his groups work on parallel I/O and specifically from the PASSION project [78, 79, 245, 42, 40]. The models provide abstractions with which a runtime system and compiler can work so that they do not have to deal with machine specific features and low-level architectural organizations. Models help the user or compiler translate (maintain) locality between various working spaces in which an I/O intensive program operates (Bordawekar [40] considers Program Space, Processor Space, File Space, and Disk Space). Three execution models; namely, the Local Placement Model, the Global Placement Model and the Partitioned In-core Model [78] can be used for designing runtime systems and compilation support.

The model based on the parallel array database introduced in Subsection 6.1.4 provides the highest abstraction level and can mainly be used for designing language and compilation support.

6.1.1 Local Placement Model

In the Local Placement Model (LPM), the out-of-core local array (OCLA) of each processor is stored in a separate *logical* file called the Local Array File (LAF) of that processor. Each LAF can be stored separately as a different physical file or different LAFs can be appended to generate a single physical file. Every processor is said to *own* its LAF.

The portion of the array which is in main memory is called the In-Core Local Array (ICLA).

The node program explicitly reads from and writes to the LAF when required. A processor cannot explicitly access a file owned by some other processor. If a processor needs to access data from a file owned by a different processor, the required data first has to be read by the owner and then communicated to the requesting processor. Same principle applies to a write or store operation. Since each LAF contains the OCLA of the corresponding processor, the distributed (or user-specified) view of the OOC global array is preserved. In other words, locality (spatial and sequential) in processor space is translated into locality in the file space.

One way to view the Local Placement Model is to think of each processor as having another level of memory which is much slower than the main memory. The extra memory is in form of a *virtual* disk which may consist of more than one physical disk. Each processor stores its LAF into its virtual disk. In other words, it is a straight-forward extension of the distributed memory model. Note that each processor may lack exclusive access to one or more physical disks that comprise its virtual disk. Disks may be shared by more than one processor. The mapping of physical to virtual disks is performed at runtime.

To store the data in the LAFs based on the distribution pattern specified in the program, redistribution of the data may be needed in the beginning when the data is staged. This is because the way data arrives (e.g., from archival storage or over the network) may not conform to the distribution specified in the program. Redistribution requires reading the data from the external storage, shuffling the data over the processors and writing the data to the local virtual disks. This increases the overall cost of data access. This cost can be amortized if an OOC array is used repeatedly.

Fig. 6.3 presents an out-of-core array distributed in **BLOCK-BLOCK** fashion over 4 processors. Each processor stores its OCLA in its LAF (shown using different shades). Each local array file is stored into a virtual disk. During computation, each processor brings the current ICLA into its memory and operates on it and stores it back (if required by the computation).

6.1.2 Global Placement Model

In the Global Placement Model (GPM), an OOC global array is stored in a single file called the Global Array File (GAF). The global array is stored in the GAF and the file is stored in a single virtual disk that is shared by all the processors accessing this array. As in case of the Local Placement Model, the virtual disk consists of one or many physical disks and the mapping to physical to virtual disks is performed at runtime.

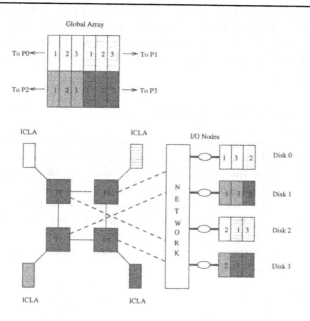

Fig. 6.4. Global Placement Model.

Each processor can read and write parts of its local array from the GAF. Since the local arrays of the processors are interleaved in the GAF, there is no concept of a local array file. A processor can *read* data owned by other processors, however, it can not *write* over the data owned by other processors. This is due to the fact that the underlying programming model is SPMD in which computations are performed in the local name space. This property of GPM eliminates any consistency problems arising from data sharing. Since any processor can read any other processor's data, the I/O can be performed in the global name space (using global coordinates). As the computation is still in the local name space, this model can be viewed as having shared memory on the files and distributed memory on the processors.

In the GPM, the global array view is preserved from a global name space HPF program. Consequently, localities from program space (temporal and spatial) are translated into localities in file space.

Fig. 6.4 shows columns of local array of a processor interleaved with others in the global file (shown with different shades). The global file is distributed over four disks which are shared by all processors. Thus, parts of the global array, based on the array distribution, need to be brought into the memory for processing. For example, Fig. 6.4 shows a sweep of the global array by columns (numbered 1 through 3), assuming that only one column per processor fits into memory. The corresponding file access pattern is also shown. In general, the accesses may be viewed as loosely synchronous. That is, the I/O can be performed in a collective manner (similar to collective communication) using a global index space. Subsequently, the data may need to be re-shuffled in memory by mapping the global indices (of current data) into local index space. Since there are no local files in the GPM, data redistribution into files is not required.

6.1.3 Partitioned In-core Model

Partitioned In-core Model (PIM) is a combination of the Global and Local Placement models. The array is stored in a single global file as in the Global Placement Model but there is a difference in the way data is accessed. In the PIM, the global array is logically divided into a number of *partitions*, each of which can fit in the combined main memory of the processors. Thus computation on each partition is primarily an in-core problem and no I/O is required during the computation of the partition.

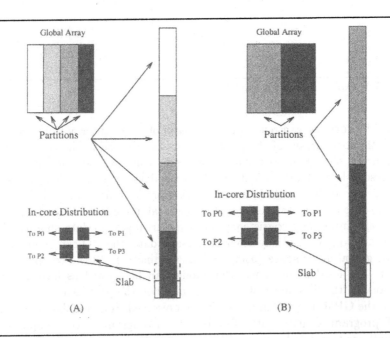

Fig. 6.5. Partitioned In-core Model.

Like LPM, PIM is also based on the data distribution concept proposed in HPF. The Local Placement Model extends the distribution concept to the out-of-core data while the Partitioned In-core Model uses the in-core distribution pattern for assigning out-of-core data to processors. In case of an in-core program, the array to be distributed is assumed to be in-core (i.e.

it can fit in combined memory of all processors). Similarly, the Partitioned In-core Model uses the combined memory to find an in-core data partition and then distributes the partition according to the user-provided distribution pattern.

PIM creates some extra complications in data ownership computations. In LPM, the out-of-core global array is distributed as if it were an in-core array. Therefore, the data ownership of the out-of-core data can be easily computed using its distribution. In case of PIM, the data ownership depends on the cumulative available memory as well as distribution pattern. Fig. 6.5 presents an out-of-core array to be distributed in a **BLOCK-BLOCK** fashion over 4 processors. Fig. 6.5 A and B illustrate in-core data distribution for two different partition sizes. Note that data belonging to processor 2 in partition (A) may lie in processor 3 in partition (B). But in both cases, the in-core array (i.e., partition) is distributed in the **BLOCK-BLOCK** fashion.

Like GPM, PIM also performs the I/O in global name space and the computation in local name space. In both models, data read and write can be easily performed using the collective methods such as the two-phase access method [38, 99]. A qualitative comparison of the LPM, GPM, and PIM models can be found in [40].

6.1.4 Model Based on a Parallel Array Database

Compilation techniques described in this chapter utilize another model of out-of-core computation which doesn't explicitly deal with files.

In this model, the compiler assumes out-of-core arrays are stored in the *parallel array database (PAD)* and it only sees the interface to procedures that operate on *PAD*.

Array sections from *PAD* are associated with the appropriate processors according to the VF data distribution specification. The data distribution specification determines the work distribution, if the owner computes rule is applied. Then the processor that a portion of out-of-core array is associated with is responsible for computing the values of elements of this array portion. All computations are performed on the data in processor's local memories. During the course of the computation, sections of the array are fetched from *PAD* into the local memory, the new values are computed and the sections updated are stored back into *PAD* if necessary. The computation is performed in stages where each stage operates on a different part of the array.

PAD is handled by procedures of the runtime system. The runtime system optimizes the data layout on disks and automatically fetches the appropriate sections from *PAD* and transfers them to the processors they asked for.

6.2 Compilation Strategy

Compiling out-of-core programs is based on the following model. All computations are performed on the data in local main memories. VFCS restructures the source program in such a way that during the course of the computation, sections of the array are fetched from disks into the local memory, the new values are computed and the sections updated are stored back to disks if necessary. The computation is performed in phases where each phase operates on a different part of the array called a *slab*. Loop iterations are partitioned so that data of slab size can be operated on each phase. Each local main memory has access to the individual slabs through a "window" referred to as the *in-core portion* of the array. VFCS gets the information which arrays are out-of-core and what is the shape and size of the corresponding in-core portions in the form of out-of-core annotation.

The process of transforming a source out-of-core program into the out-of-core SPMD program can be conceptually divided into five major steps:

1. *Distribution of each out-of-core array among the available processors.* Array elements that are assigned to a processor according to the data distribution are initially stored on disks. Further, the resulting mapping determines the work distribution. Based on the out-of-core annotation, memory for in-core array portions is allocated.

2. *Distribution of the computation among the processors.* The work distribution step determines for each processor the *execution set*, i.e. the set of loop iterations to be executed by this processor. The main criteria is to operate on data that are associated with the "nearest" disks and to optimize the load balance. In most cases the "owner-computes-rule" strategy is applied; the processor which owns the data element that is updated in this iteration will perform the computation.

3. *Splitting execution sets into tiles.* The computation assigned to a processor is performed in phases called *tiles* where each phase operates on a different slab. Loop iterations are partitioned so that data of slab size can be processed on each phase.

4. *Insertion of I/O and communication statements.* Depending on the data and work distribution, determine whether the data needed is in the local or remote in-core portion or on a disk and then detect the type of communication and I/O operation required.

5. *Organization of Efficient Software Controlled Prefetching.* Efficiency of parallel I/O accesses can be dramatically increased if the compiler and runtime system are able to prefetch data to disk buffers or disks from which the transfer to the target processors is most optimal.
   In Section 6.6, we deal with two methods which support the I/O oriented prefetching. The first method is based on the compile time program analysis which determines the placement of data prefetching directives in the source code. In the second method, the compiler passes information

about accesses to out-of-core structures and predicted execution time in the form of an annotated program graph, referred to as Input-Output Requirement Graph (IORG), to the runtime system. This graph is incrementally constructed in the program database of the VFCS during the compilation process and written to a file at its end. IORG is used by the runtime system in the optimization of prefetching.

The basic out-of-core compilation techniques were elaborated by the Syracuse group [45, 244].

6.3 Specification of the Application Area

Codes considered in this chapter include the following classes:

- *Regular computations*
  Regular codes can be precisely characterized at compile time. We distinguish between loosely synchronous and pipelined computations.
  - *Loosely synchronous computations*
    Loosely synchronous problems ([115]) are divided into cycles by some parameter which can be used to synchronize the different processors working on the problem. This computational parameter corresponds to e.g., time in physical simulations and iteration count in the solution of systems of equations by relaxation. In such problems, it is only necessary to identify and insert efficient vector or collective communication at appropriate points in the program[3]. The Jacobi code in Fig. 6.6 is a typical representative of this class.
  - *Pipelined computations*
    A different class of computations contains *wavefront parallelism* because of existing loop-carried cross-processor data dependences that sequentialize computations over distributed array dimensions. To exploit this kind of parallelism, it is necessary to pipeline the computation and communication [222, 248].
- *Irregular computations*
  In irregular codes, major data arrays are accessed indirectly, usually via some form of index array. The underlying application domain is generally either sparse or unstructured. Application areas in which such codes are found range from unstructured multigrid computation fluid dynamic (CFD) solvers, through molecular dynamics (MD) codes and diagonal or polynomial preconditioned iterative linear solvers, to time dependent flame modeling codes [52].

In the first part of this chapter we apply out-of-core transformations to regular and in the second part to irregular codes.

---

[3] Such codes are sometime referred to as fully data-parallel codes.

6.4 Compiling Out-of-Core Regular Codes

6.4.1 Out-of-Core Restructuring Techniques Applied to Loosely Synchronous Computations

We illustrate the application of our out-of-core transformation techniques on the Jacobi code in Fig. 6.6.

```
PARAMETER :: M = ..., N = ..., IWIDTH = ..., ITIME = ...
PROCESSORS  PROC(M,M)
REAL,  DISTRIBUTED ( BLOCK , BLOCK ),  OUT_OF_CORE, &
                 IN_MEM (:,IWIDTH) :: A(N,N), B(N,N)
          ...
DO   IT = 1, ITIME
      DO   J = 2, N-1
           DO   I = 2, N-1
                A(I,J) = 0.25 * (B(I-1,J) + B(I+1,J) + B(I,J-1) + B(I,J+1))
           END DO
      END DO
      ...
END DO
```

Fig. 6.6. Jacobi Program.

Because the out-of-core method is an extension of the appropriate in-core one, we introduce the in-core method first; we repeat some terms already introduced in Section 4.2.1.

In Fig. 6.7, we reproduce the in-core Jacobi code, transformed to run on a set of $M \times M$ processors, using a pseudo notation for message passing code. We have simplified matters by only considering the code for a non-boundary processor. Further, we assume that the array sizes are multiples of M. Optimization of communication has been performed insomuch as messages have been extracted from the loops and organized into vectors for sending and receiving. Each processor has been assigned a square subblock of the original arrays. The compiler has declared local space of the appropriate size for each array on every processor. Array B has been declared in such a way that space is reserved not only for the local array elements, but also for those which are used in local computations, but are actually owned by other processors. This extra space surrounding the local elements is known as the overlap area. Values of A on the local boundaries require elements of B stored non-locally for their computation. These must be received, and values from local boundaries must be sent to the processors which need them. The work is distributed according to the data distribution: computations which define the data elements owned by a processor are performed by it - this is known

---

/* code for a non-boundary processor p having coordinates (P1,P2) /*

PARAMETER(M = ..., N = ..., LEN = N/M)

/* declare local arrays together with overlap areas */
REAL A(1:LEN,1:LEN), F(1:LEN,1:LEN), B(0:LEN+1,0:LEN+1)

/* global to local address conversion */
/* $\widehat{f(I)}$ represents: f(I) – $L1(P1) + 1 and $\widehat{g(J)}$ represents: g(J) – $L2(P2) + 1 */
    ...
DO ITIME = 1, ITIMEMAX

/* send data to other processors */
  SEND (B(1,1:LEN)) TO PROC(P1-1,P2)
  SEND (B(LEN,1:LEN)) TO PROC(P1+1,P2)
  SEND (B(1:LEN,1)) TO PROC(P1,P2-1)
  SEND (B(1:LEN,LEN)) TO PROC(P1,P2+1)

/* receive data from other processors, assign to overlap areas in array B */
  RECEIVE B(0,1:LEN) FROM PROC(P1-1,P2)
  RECEIVE B(LEN+1,1:LEN) FROM PROC(P1+1,P2)
  RECEIVE B(1:LEN,0) FROM PROC(P1,P2-1)
  RECEIVE B(1:LEN,LEN+1) FROM PROC(P1,P2+1)

/* compute new values on local data λ(p) = A($L1(P1):$U1(P1),$L2(P2):$U2(P2))
*/
  DO I = $L1(P1), $U1(P1)
    DO J = $L2(P2), $U2(P2)
       A($\widehat{I}$, $\widehat{J}$) = 0.25*(F($\widehat{I}$, $\widehat{J}$)+ &
                   B($\widehat{I-1}$, $\widehat{J}$)+B($\widehat{I+1}$, $\widehat{J}$)+ B($\widehat{I}$, $\widehat{J-1}$)+B($\widehat{I}$, $\widehat{J+1}$))
       END DO
    END DO
    ...
END DO

Fig. 6.7. SPMD Form of In-Core Jacobi Relaxation Code for a Non-Boundary
Processor.

as the *owner computes* paradigm. The processors then execute essentially
the same code in parallel, each on the data stored locally. We preserve the
global loop indices in the SPMD program. Therefore, the global to local
index conversion must be provided to access the appropriate local elements
of A and B. The iteration ranges are expressed in terms of the *distribution
(local) segment* ranges (see Section 4.2.1) which are parameterized in terms
of the executing processor coordinates. It is assumed that if processor p has
coordinates *(P1,P2)* then the local segment of A on processor p, $\lambda^A(p) =$
A($L1(P1):$U1(P1), $L2(P2):$U2(P2)).

```
PARAMETER:: M = ..., N = ..., IWIDTH = ..., LEN = N/M, ITILE = IWIDTH
TYPE (DD) :: dd_A, dd_B        ! distribution descriptors
TYPE TRIPLET                   ! description of one dimension of an ooc section
      INTEGER :: l,u,s
END TYPE TRIPLET
TYPE (TRIPLET), DIMENSION (:), POINTER :: secsin, secsout  ! triplet lists

/* declare local in-core portions */

REAL A(1:LEN,1:IWIDTH), B(0:LEN+1,0:IWIDTH+1)

/*  global to local index conversion */
/*  f(I) represents: f(I) - $L1(P1)+1 and   */
/*  g(J) represents: g(J)-($L2(P2)+(K-1)*IWIDTH)+1 */


       . . .
ALLOCATE (secsin(2), secsout(2))   ! descriptors for two-dimensional sections
secsin(1) = TRIPLET($L1(p)-1, $U1(p)+1, 1)
secsout(1) = TRIPLET($L1(p), $U1(p), 1)

DO IT = 1, ITIME
    DO  JJ = MAX(2,$L2(P2)), MIN(N-1, $U2(P2)), ITILE
        JLWB = JJ
        JUPB = MIN(JJ+ITILE-1, N-1, $U2(P2))
        secsin(2) = TRIPLET(JLWB-1, JUPB+1, 1)
        CALL  VIPIOS_Get_Sec(B,dd_B,secsin)
        DO  J = JLWB, JUPB
            DO I =  $L1(P1)), $U1(P1)
               A(I, J) =
                        0.25*(B(I-1, J)+B(I+1, J)+B(I, J-1)+B(I, J+1))
            END DO
        END DO
        secsout(2) = TRIPLET(JLWB, JUPB, 1)
        CALL  VIPIOS_Put_Sec(A,dd_A,secsout)
    END DO
        . . .
END DO
```

Fig. 6.8. SPMD Form of Out-of-Core Jacobi Relaxation Code.

```
interface
  subroutine VIPIOS_Get_Sec (A,dd_A,triplet_list)
    integer,dimension(:,:),intent(in) :: A              ! address of ICP
    type(dist_desc),intent(in):: dd_A                   ! distribution descriptor for A
    type(triplet),dimension(:),pointer:: triplet_list   ! section description
  end subroutine VIPIOS_Get_Sec
end interface
```

Fig. 6.9. Interface for VIPIOS_Get_Sec

In the out-of-core version[4] (Fig. 6.8), at any time, only a portion of B is fetched from the parallel array database (PAD) and stored in processor's memory. The portion of B that is currently in processor's memory is called the In-Core Portion (ICP). Similarly, the new values are computed for the ICP of A and this ICP is stored back into appropriate locations in the PAD. The size of the ICP is derived from the IN_MEM specification and the number of available processors. In the code, one can see that the larger the ICP the better, as it reduces the number of I/O operations. The computation of new values for the entire local segment of A is performed in stages where each stage updates a different part of the array called a *slab*. The size of each slab is less or equal to the size of the ICP. There are two levels of data and iteration set partitioning: data is first partitioned among processors and then data within a processor is partitioned into slabs which fit in the processor's local memory; iterations are first partitioned among processors applying the owner computes rule that relies on the data partitioning and the local iterations are further partitioned into tiles [70, 266, 267, 265, 268] in such a way that they operate on the appropriate slab data.

```
INTEGER, PARAMETER :: C = ..., R = ...
PROCESSORS P(4)
REAL, DISTRIBUTED (:,BLOCK), OUT_OF_CORE, IN_MEM (R,C) :: A(100,10000)
    ...
DO IT = 1, ITIME
  DO I = 1, 100
    DO J = 2, 10000
      A(I,J) = F(A(I,J-1))
    END DO
  END DO
END DO
```

Fig. 6.10. Vienna Fortran OOC Program - a Candidate for Pipeline Execution

---

[4] In the PARAMETER parts, we assume that all dividends are multiples of the corresponding divisors.

The appropriate sections of B are fetched from the PAD by the VIPIOS procedure VIPIOS_Get_Sec whose interface is specified in Fig. 6.9. Sections of A are written to the PAD by the procedure VIPIOS_Put_Sec. Its arguments have the same meaning as in case of VIPIOS_Get_Sec. Structure dd_A stores the distribution descriptor associated with A; it includes information about: shape, alignment and distribution, associated processor array, size of the local data segment of A and size of the IP of A in processor p. In this example, each section is described by a two-element list of triplets.

Remark: The interface in Fig. 6.9 serves for the illustration purpose only. In the real implementation, VIPIOS_Get_Sec and VIPIOS_Put_Sec will provide generic interface supporting general regular section descriptions.

6.4.2 Out-of-Core Restructuring Techniques Applied to Pipelined Computations

The access pattern leading to pipelining can be found e.g. in the Gauss-Seidel program, where parallelism is along a wavefront. Communication and computation have to be pipelined in order to achieve the best tradeoff between minimizing the number of messages and exploiting parallelism. The SPMD program version achieves this by sending the array elements in blocks, a compromise between sending them one at a time and sending them all at once. The importance of this optimization was demonstrated in [222, 248]. The transformation necessary to achieve this is strip mining. In Fig. 6.10 there is the Vienna Fortran program that is a candidate for the pipeline implementation. Its in-core and out-of-core SPMD forms are presented and described in [60].

6.4.3 Optimizations for Out-of-Core Regular Codes

So far, we have introduced out-of-core program transformations without any I/O optimizations performed. Notice that this translation is a *straightforward extension* of in-core compilation strategy.

The basic approach outlined in the previous subsection does not consider optimizations like minimizing file access costs, determining tile shape, optimal tile scheduling, and prefetching and caching data. Several optimizations are briefly discussed in this subsection. Prefetching issues are in general discussed in Section 6.6.

From the point of view of data access costs incurred during the execution of an OOC program, a *three level memory hierarchy model* can be considered which is depicted in Fig. 6.11. It is assumed that the processors reference their *local primary memories* in unit time, $t_1 = 1$. The model further stipulates that a processor's reference to data that is not in its own internal memory requires a latency of t_2 time units fulfilled if the data is currently in memory

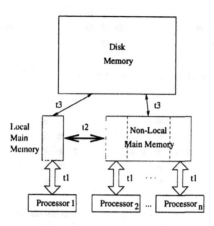

Fig. 6.11. Memory Hierarchy.

of a remote processor and t_3 time units if the data is in disk memory. It is assumed that $t_3 \gg t_2 \gg t_1$. These memory hierarchy features force programmers and compiler developers to attend to the matter of

- locality and data reusing, i.e. how to arrange computations so references to data in local internal memory are maximized, and the time-consuming references to data on disks are minimized;
- hiding communication and I/O latency by computations.

Eliminating Extra File I/O by Reordering Computation. In this paragraph, we focus on optimization of out-of-core stencil computations.In this context, we briefly discuss an optimization method proposed by Bordawekar et al. [44].

There are many scientific and engineering problems in which the major computation structure is regular. This kind of regularity is a great advantage, contributing to the good performance of many parallel implementations.

The regular computation model involves repetitive evaluation of values at each mesh point with local communication. The computational workload and the communication pattern are the same at each mesh point. This class of computations naturally arises in numerical solutions of partial differential equations and simulations of cellular automata.

The numerical solution of partial differential equations, by methods such as point Jacobi iteration, involves evaluation of the value at each mesh point at each iteration as the weighted sum of previous values of its neighbors. The pattern of communicating neighbors is called the *stencil* [179]. Fig. 6.12 shows several commonly used stencils for two-dimensional meshes.

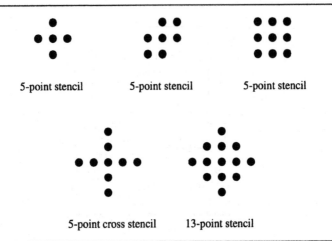

Fig. 6.12. Commonly Used Stencils

The Jacobi code which we several times used for the illustration of compilation techniques represents the 5-point stencil computation. On the other hand, Fig. 6.13 presents a Vienna Fortran program performing a 9-point star stencil relaxation over an out-of-core array A which is distributed in **BLOCK** fashion in both dimensions. The underlying processor grid consists of nine processors arranged as a 3 x 3 grid (Fig. 6.14).

```
PARAMETER :: N = ...
PROCESSORS PROC(3,3)
REAL, OUT_OF_CORE, DISTRIBUTED ( BLOCK, BLOCK ) :: A(N,N), A_new(N,N)
...
DO  I = 2, N-1
    DO  J = 2, N-1
        A_new (I,J) = (A(I,J-1)+A(I,J+1)+A(I+1,J)+A(I-1,J) &
                      +A(I+1,J+1)+A(I+1,J-1)+A(I-1,J-1)+A(I-1,J+1))/8
    END DO
END DO
...
```

Fig. 6.13. A 9-Point Relaxation Over an OOC Array A.

Before analyzing communication in OOC stencils problems, let us briefly review communication patterns in in-core stencil problems. Let us assume that each processor can store its out-of-core local array (OCLA) in its memory. Fig. 6.14 illustrates the data which is to be fetched by processor 5 from its neighbors and which is required for computation of its entire OCLA (overlap

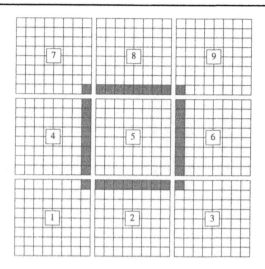

Fig. 6.14. In-Core Communication Patterns for the Stencil

area). Observing Fig. 6.14 one can easily notice that every processor (other than those at the boundary) has to communicate the overlap data with its eight neighbors; this communication pattern can be easily implemented by exchanging data between processor pairs (see Section 4.2).

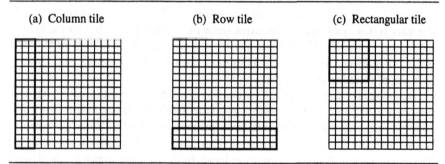

(a) Column tile   (b) Row tile   (c) Rectangular tile

Fig. 6.15. Tile Shapes

Remember that an ICLA (or tile) is a section of the out-of-core array which can fit in the in-core memory. Shape of a tile is determined by the extent of the tile in each dimension. Fig. 6.15 depicts a column tile, a row tile, and a square tile.

Fig. 6.16 shows overlap areas for two processors for the column version of the ICLA. The overlap area is colored using two shades. The overlap area denoted by black represents that data organized on disk by the VIPIOS in

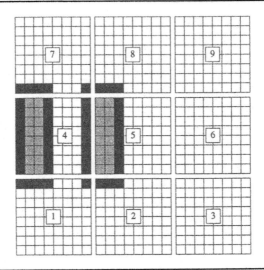

Fig. 6.16. Out-of-Core Communication Patterns for the Stencil

such a way, that it can be efficiently transferred to the respective processor; the data can be stored on local disk of this processor.

On the other hand, the dark shaded overlap areas represent the data that are organized on disk by VIPIOS, due to the data distribution specification, in a way that results in efficient transfer to the processors owning the segments which include these data.

In Fig. 6.16, processors 4 and 5 are both using their first (shaded) ICLA. For this ICLA, processor 4 requires data from its top and bottom neighbor (processor 1 and 7). Since these processors will also be working on the first ICLA, they have the necessary boundary data in their memory and there is no need of fetching it from disk.

However, for the first ICLA, processor 5 needs to communicate with processors 1,2,4,6, and 8. Processor 5 needs to fetch the last column from processor 4 and the corner elements from processors 0 and 6. Since processor 3 is also working on the first ICLA, it does not have this data in its memory. Hence, processor 4 needs to read column from its local array file and send it to processor 5. This results in processor 5 performing extra file access. Similarly, extra I/O accesses are performed by processors 1 and 7. Moreover, processor 5 has to wait until processors 4, 1, and 7 have read and sent it to processor 5. This also results in extra overhead.

Bordawekar et al. [44] propose a method (a *cookie-cutter* approach) for eliminating the extra file accesses when using the local placement model. This method can be directly applied when using the Parallel Array Database Model (subsection 6.1.4).

When applying this method to our example (Fig. 6.13) the resulting tile scheduling for column and square tiles will have the form depicted in Fig.6.17.

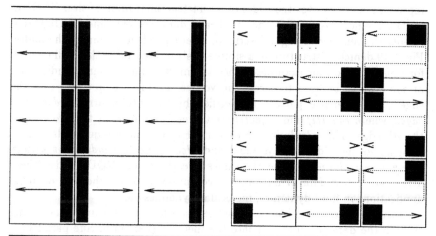

Fig. 6.17. Tile Scheduling

6.5 Parallelization Methods for Out-of-Core Irregular Problems

Large scale irregular applications[5] involve large arrays and other data structures. Runtime preprocessing and automatic partitioning provided for these applications results in construction of large data structures which increase the memory usage of the program substantially. Consequently, a parallel program may quickly run out of memory. Therefore, the development of appropriate parallelization methods for out-of-core irregular programs is relevant and important research issue. Even if, the performance is somewhat worse than the case when the same size problem could fit in the memory, the capability to solve larger problems provides the appropriate flexibility to the users. Furthermore, not all users have access to large-scale systems with huge amounts of memory. Providing support that facilitates out-of-core solutions will enable such users to solve relatively larger size problems[6].

In the following sections, we describe compilation and runtime techniques to support out-of-core irregular problems.

As described in Sections 4.3 and 6.3, irregular problems are characterized by indirect data accesses where access patterns are only known at runtime.

---

[5] Most real world applications are large-scale, but their sizes have been limited by the memory of the available systems.

[6] This is also known as the scale-down problem.

The input data (e.g., a mesh representing the discretization of an air-foil) represents a description of the problem domain which is irregular.

Fig. 6.18 illustrates a typical irregular loop. The loop L2 in this code represents a sweep over the edges of an unstructured mesh of size: NNODE, NEDGE, where NNODE is the number of data elements and NEDGE represents the number of edges describing interdependencies among the nodes. Fig. 6.19 depicts a structure of a simple unstructured mesh that will be used for illustration of our examples. Since the mesh is unstructured, indirection arrays have to be used to access the vertices during a loop sweep over the edges. The reference pattern is specified by the integer arrays *edge1* and *edge2*, where *edge1(i)* and *edge2(i)* are the node numbers at the two ends of the *i*th edge. The calculation on each node of the mesh requires data from its neighboring nodes. Arrays *edge1* and *edge2* are called *indirection arrays*. The arrays *x* and *y* represent the values at each of the NNODE nodes; these arrays are called *data arrays*. Such a computation forms the core of many applications in fluid dynamics, molecular dynamics etc.

A typical parallelization based on the inspector-executor model, assuming that all the data structures fit into the memory (i.e., an in-core parallelization) process involves several steps; namely, 1) Data Distribution, 2) Work Distribution, 3) Inspector, and 4) Executor. The primary goal of these steps and techniques is to minimize communication while obtaining load balance so that the a good parallelization is obtained. Even when the data is out-of-core, primary goals remain the same, that of minimizing various overheads. However, since I/O is very expensive, one of the most important (and probably more important than the others) goals is to minimize I/O accesses. Therefore, in order to extend the inspector-executor model, I/O access costs must be accounted for and incorporated into various steps.

As the models described in the previous sections indicate, a computation step in any OOC computation will require reading data from disks in slabs (or tiles), processing the data, and writing it back (if necessary). Therefore, it is important to minimize such steps. In other words, the steps described above must be modified so that the data is reorganized on disks at runtime so as to minimize I/O accesses. Furthermore, scheduling of slabs in the memory must be performed in such a way that not only the data present in the local memory is maximally used before being discarded (or written back onto disks), but the data, if required by other processors, should also be used during this phase.

In the following, we propose several techniques which address these problems.

6.5.1 Problem Description and Assumptions

We assume that the size of the data, data structures describing interactions among data and those describing computation patterns (e.g., indirection arrays, neighbor lists etc.) are very large. However, we assume that the data can

```
         PARAMETER :: NNODE = ..., NEDGE = ..., NSTEP = ...
         REAL x(NNODE), y(NNODE)                    ! data arrays
         INTEGER edge1(NEDGE), edge2(NEDGE)         ! indirection arrays

L1:      DO j = 1, NSTEP                            ! outer loop
L2:          DO i = 1, NEDGE                        ! inner loop
                 x(edge1(i)) = x(edge1(i)) + y(edge2(i))
                 x(edge2(i)) = x(edge2(i)) + y(edge1(i))
             END DO
         END DO
```

Fig. 6.18. An Example with an Irregular Loop.

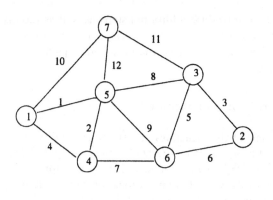

i	edge1(i)	edge2(i)
1	1	5
2	5	4
3	2	3
4	1	4
5	5	2
6	2	6
7	6	4
8	5	3
9	6	5
10	1	7
11	7	3
12	5	7

Fig. 6.19. Example Mesh with Seven Nodes and Twelve Edges.

fit in the system's memory, while the data structures describing interactions etc. are out-of-core. This assumption is realistic in many applications where the data size is N and there are few arrays storing the data, but there are a large number of structures describing the interactions with sizes of $K * N$ (e.g., CFD) or $O(N^2)$ (e.g., molecular dynamics). For example, if the value of K is 10 and there are 5 such data structures, then the amount of memory required to run an application with data size of N will be at least $50N$ without considering temporary storage requirements. Therefore, even if the data structures containing the primary data of a problem of size N could fit into memory, data structures describing interactions are still out-of-core.

In the next subsection, we consider the parallelization strategy for this restricted problem.

6.5.2 Data Arrays are In-Core and Indirection/Interaction Arrays Out-of-Core

This class of problems will be referred to as the *DAI/IAO* (*Data Arrays are In*-core / *Indirection Arrays are Out*-of-core) problems.

We describe our parallelization scheme using loop *L2* shown in Fig. 6.18 and the unstructured mesh (with NNODE=7 and NEDGE=12) depicted in Fig. 6.18. We only consider the Local Placement Model.

The main goal of the proposed technique is to minimize I/O costs during the execution of an application by data reorganization on disks and by using efficient schedules to stage data into memory. Specifically, we want to minimize I/O for both reading/writing slabs as well as for communication[7]. Therefore, our initial goal is to introduce additional steps as well as modify existing steps in the inspector-executor model to satisfy the goals of minimizing I/O costs. In the following, we describe the proposed steps using the kernel illustrated in Fig. 6.19.

Parallelization Steps and Runtime Support

This part presents an overview of the principles and the functionality of the extended CHAOS runtime library, called *CHAOS/E*, that our parallelization scheme is based on. CHAOS/E can be used either by the application programmers to parallelize the *DAI/IAO* applications, or can be used by parallelizing compilers to automatically generate parallel OOC code for the *DAI/IAO* applications.

The *DAI/IAO* computation is performed in 7 steps that are outlined below.

A. Default Initial Data and Work Distribution
B. Partition Data Arrays
C. Redistribute Data Arrays
D. Compute new iteration tiles
E. Redistribute local files with indirection arrays
F. Out-of-Core Inspector
G. Out-of-Core Executor

Each of these steps is described in more detail below.

---

[7] Note that since data sets reside on disks, even the communication steps potentially require I/O if the data to be communicated is not present in the sender's memory at the time when communication takes place.

Processor 1 (P1)   Processor 2 (P2)

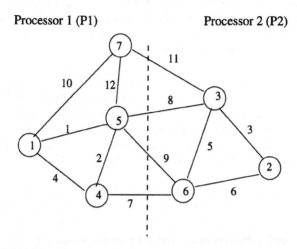

Fig. 6.20. Example Mesh with Seven Nodes and Twelve Edges Distributed on 2 Processors.

A. Clearly, in order to fully exploit parallelism, the step of data partitioning must also be done in parallel. Initially, it is assumed that the iterations and data arrays are distributed among processors in a known uniform manner. Thus, the global iteration set is partitioned blockwise among the processors and the iteration set assigned to a processor is further blockwise split into a number of tiles. Section of an indirect array associated with each tile can fit in the main memory of any one of processors.

B. This step essentially involves the use of a partitioner which uses out-of-core data to determine a new data distribution to minimize interprocessor communication and enhance locality and load balance. Note that any partitioner, traditionally used for in-core data, can be extended to be used in this situation. Partitioners can be available as library functions invoked by a user/compiler. A partitioner returns an irregular assignment of array elements to processors, which are stored as another data structure called *translation table*. A translation table lists the owner processor and the logical offset address of each data array element.

There are many partitioning heuristics methods available based on physical phenomena and proximity. Partitioners (such as available in CHAOS or their appropriate extensions) are linked to programs by using data structures that store information on which data partitioning is to be based. They can utilize spatial, load, or connectivity information or combination of them. For example, a code segment can be inserted into the SPMD program to construct a graph representing the access patterns of the distributed arrays in the loops. The graph is constructed by executing a modified version of the loop which forms a list of edges. Note that the operations in the loop are not performed. This list is stored on disks. The generated graph is passed to a graph partitioner. The graph partitioner returns the new array distributions.

C. In step C, the data arrays that were initially distributed in block fashion are remapped to obtain the new distribution determined by the partitioner. For example, Fig. 6.20 shows a possible data distribution of the nodes of our example mesh across 2 processors. This would be the the result of redistribution from an initial distribution in which Processor 1 had nodes 1-4 and Processor 2 had nodes 5-7.

D. Once the new array distribution is obtained, loops iterations that access the distributed arrays must also be partitioned among processors to balance computational load and maintain locality (to reduce communication costs). Furthermore, the indirection arrays must be reorganized (within each processor), to minimize I/O costs. To partition the loop iterations, for locality and load balance, the primitives used here generate a processor assignment graph for each iteration [214]. A loop iteration is assigned to a processor that owns the maximum number of distributed array elements referenced in that iteration.

In the second part of this step, the data is reorganized within each local file to minimize I/O. The iteration set assigned to a processor is further split into a number of tiles/slabs, each large enough to enable all the relevant data to fit in the memory. The tiles are reorganized and are classified into two groups; namely, *local tiles* and *non-local tiles*. The local tiles include only those iterations for which the corresponding data is available locally (i.e., on local files). For example, iterations 1, 2, 4, 10, and 12 in Fig. 6.20 will be part of local tiles. This reorganization and classification enables creating large tiles because no buffers are required for communication thereby reducing the number of I/O operations. Secondly, this grouping allows prefetching and pipelining of local tiles because the access is independent of other processors since there is no communication. Therefore, computations can be overlapped with I/O. That is, not only the number of I/O operations may be reduced, but I/Os can be overlapped with computation. The non-local tiles include iterations that require off-processor accesses. For example, iterations 7, 8, 9, and 11 in Fig. 6.20 require off-processor accesses. This grouping is also advantageous because communication may be coalesced for each title, and I/O required for communication may be reduced by coalescing them together. The primitive iter_partitioner performs the iteration partitioning and returns a descriptor that specifies the tiles to be executed on each processor.

E. In this step, the local files containing indirection arrays are remapped to conform with the loop iteration partitioning in the previous step.

F. Step F represents the inspector. However, since the inspector requires building communication schedules for only non-local tiles, it does not require a pass through the entire local data set. Two options for building communication schedules are available. In the first option, a communication schedule is generated for each non-local tile, whereas in the second option, communication schedules for all non-local tiles may be combined. The former allows overlapping of computation with communication to remote processors, while the latter option reduces the number of communication steps.

G. Phase G represents the executor. In this phase, for each time step, the following steps are performed. 1) Fetch local tiles and update them, 2) Gather non-local data, 3) Update non-local tiles, and 4) Scatter data to update non-local tiles on remote processors. Note that steps 1 and 2 may be overlapped. Similarly, steps 3 and 4 may be overlapped if the second communication option in step F is chosen.

Implementation Design

All the processing phases discussed above are illustrated by means of the running example shown in Fig. 6.18, where NNODE=7 and NEDGE=12. Code implementing the phases is shown in Fig.s 6.22, 6.24, 6.25, and 6.26 and discussed in the following paragraphs.

Data Partitioning. We discuss the application of two partitioners:

1. *recursive spectral bisection partitioner* which is a graph partitioner that utilizes connectivity information
2. *recursive coordinate partitioner* whose partitioning algorithm is based on spatial coordinate information.

Coupling a Graph Partitioner. Some data partitioners determine data distribution by breaking up a graph, where the nodes of the graph represent the data array elements and the edges of the graph depict the dependency pattern between the nodes. The graph partitioners divide the graph into equal subgraphs with as few edges as possible between them.

To interface with graph partitioners, a distributed data structure called Runtime Data Graph (RDG) must be constructed. Input data for the construction of an RDG are computed by the parallel execution of a modified version of the DO loop. Therefore, loop iterations of the loop are initially distributed among processors in some uniform manner. We use the block distribution in our implementation. Processing steps needed to derive the distribution of arrays are depicted in Fig. 6.22.

Processor 1	Processor 2
lwbs = {1, 7}	lwbs = {4, 10}
upbs = {3, 9}	upbs = {6, 12}
This implicitly assumes:	This implicitly assumes:
execution set={1,2,3,7,8,9}	execution set={4,5,6,10,11,12}
tile1={1,2,3}, tile2={7,8,9}	tile1={4,5,6}, tile2={10,11,12}
local_ind_list = {1,2,3,4}	local_ind_list = {5,6,7}
n_local = 4	n_local = 3

Fig. 6.21. An Example of Initial Iteration Partitioning.

RDG is generated in 4 steps:

1. *Distribute data and loop iterations to processors in uniform blocks (in-core processing).*
   First, the iterations are distributed blockwise and then each iteration block assigned to a processor is split into tiles. This functionality is provided by the primitive build_block_tiles. It accepts the lower and upper bounds of the loop iteration space and number of tiles the execution set of each processor owns as input parameters. It returns the description of tiles to be executed on the calling processor. An example of a possible initial loop and data partitioning for the loop from Fig. 6.18 is shown in Fig. 6.21. We assume two processors and two iteration tiles are generated for each processor. The values of *lwbs* and *upbs* point to the sections of arrays *edge1* and *edge2* that are stored in file *'mesh.dat'*.

2. *Generate local RDG on each processor (out-of-core processing).*
   The RDG is generated based on the access patterns of the arrays *x* and *y* in the DO loop. A slice of the original DO loop is executed in parallel to form out-of-core local lists of array access patterns for each processor according to the work distribution. The lists are incrementally constructed executing a sequence of tiles (two tile per processor in our example). Slabs of indirection arrays are read from the global file *'mesh.dat'* and slabs of computed lists are written to the appropriate local files. The lists *lhs_ind* and *rhs_ind* record the left hand side and right hand side array access patterns in statement S1. The RDG obtained using *S1* does not get altered by the statement *S2* as there are no new pairs of indices otherwise an additional pair of lists should be constructed for *S2*. All processors then call the primitive build_local_RDG with the files storing the left hand side and right hand side array indices as input arguments. The RDG is stored in the hash table which is identified by an integer (*hash_table*).

```
OPEN (mesh_file, file ='/pfs/mesh.dat')
... initialization of x and y ...

C-Constructing iteration tiles (block partitioning)
    CALL  build_block_tiles(1, nedge, number_of_tiles, lwbs, upbs)

C-DATA PARTITIONING USING A GRAPH PARTITIONER

C-Constructing local RDG

lhs_pntr = 1; rhs_pntr = 1
DO  tile = 1, number_of_tiles
    CALL  read_slab (mesh_file,lwbs(tile),upbs(tile),edge1)
    CALL  read_slab (mesh_file,NEDGE+lwbs(tile),NEDGE+upbs(tile),edge2)
    lcount = 1; rcount = 1
    DO  i=1, upbs(tile) - lwbs(tile) + 1
        lhs_ind(lcount) = edge1(i); rhs_ind(rcount) = edge2(i)
        lcount = lcount + 1; rcount = rcount + 1
    END DO
    CALL  write_slab(lhs_ind_file, lhs_ind, lhs_pntr, lcount-1)
    CALL  write_slab(rhs_ind_file, rhs_ind, rhs_pntr, rcount-1)
    lhs_pntr = lhs_pntr + lcount-1; rhs_pntr = rhs_pntr + rcount-1
END DO

C-Initialize hash table to store RDG
    hash_table = init_ht(init_argument)
    CALL  build_local_RDG(hash_table, lhs_ind_file, rhs_ind_file)

C-Generating distributed RDG: stored in csrp and csrc
    CALL  generate_RDG(hash_table, local_ind_list, n_local, csrp, csrc, ncols)

C-Parallel RSB partitioner: returns translation table dtrat
    CALL  partitioner_RSB(local_ind_list, n_local, csrp, csrc, dtrat)
```

Fig. 6.22. Data Partitioning

3. *Merge local graphs to produce a distributed graph (in-core processing).*
In the next step, the local graphs are merged to form a distributed graph.
Procedure **generate_RDG** is called on each processor to combine the local
adjacency lists in the hash table into a complete graph (closely related
to compressed sparse row format). The RDG is distributed so that each
processor stores the adjacency lists for a subset of the array elements.
Edges are grouped for each array element together and then stored in
the list represented by the integer array **csrc**. A list of pointers, represented by the integer array **csrp**, identifies the beginning of the edge list
for each array element in **csrc**.

4. *Data Partitioning.*

The RDG representation of the distributed array access patterns is passed to the procedure partitioner_RSB. This procedure uses a data mapping heuristic. Any graph partitioner can be used here as long as the partitioner recognizes the input data structure. The parallel partitioner returns a pointer to a translation table (*dtrat*) which describes the new array distribution. A translation table describes the distributed array layout amog processors.

Coupling a Coordinate Partitioner. Fig. 6.23 illustrates a call to the recursive coordinate partitioner. Here, the use of geometric information is shown. Arrays coord1 a coord2 carry the spatial coordinates for elements in x and y.

C–DATA PARTITIONING USING RCB

C–Parallel RCB partitioner: it returns translation table dtrat
C– coord1 a coord2 are coordinate arrays

      CALL partitioner_RCB(2, coord1, coord2, ndata, dtrat)

Fig. 6.23. Partitioning that uses Two-Dimensional Geometry Information

---

C–REDISTRIBUTING DATA

C–Constructing schedules for redistribution

      CALL remap(dtrat,local_ind_list,dsched,newlindlist,newlindlist_size)

C–Redistributing data arrays

      CALL Gather(x, x, dsched)
      CALL Gather(y, y, dsched)

Fig. 6.24. Redistribution of Data

Data Remapping. Once the new array distribution has been specified data arrays must be remapped and redistributed based on it. This is done in two steps which are shown in Fig. 6.24.

1. Procedure remap is called which inputs the translation table (*dtrat*) describing the new array distribution and a list of initial array indices (*lo-*

cal_ind_list). It returns a schedule pointer (*dsched*) and a new local index list (*newlindlist*),

2. the schedule pointer is passed to the routine Gather to move data arrays *x* and *y* between the source and target distributions. The Gather routine accepts the following arguments: addresses of the local data segment and communication buffer (they have to equal if arrays are to be remapped), and a pointer to the scheduler.

---

files(1) = lhs_ind_file; files(2) = rhs_ind_file; number_of_files = 2

```
C–ITERATION PARTITIONING
C–Global references for each local iteration are stored in the files. The pointers
C–to these files are stored in the array files
C–Iterations assigned to a processor are partitioned into local and non-local tiles
C–Local files storing indirection arrays are generated; the ordering of array
C–elements is aligned with tiles
C–Iteration partitioner returns the tile description and
C–local files storing indirection arrays
C–The pointers to the local files are returned in the array files

      CALL iter_partitioner(dtrat, files, number_of_files, number_of_all_tiles, &
          number_of_local_tiles, lwbs, upbs)
```

Fig. 6.25. Iteration Partitioning

Partitioning and Redistribution of Loop Iterations. The newly specified data array distribution is used to decide how loop iterations are to be distributed among processors. This calculation takes into account the processor assignment of the distributed array elements accessed in each iteration. A loop iteration is assigned to the processor that owns the maximum number of distributed array elements referenced in that iteration.

Distributing loop iterations is done via procedure iter_partitioner which includes the generating the runtime data structure, called *Runtime Iteration Processor Assignment graph* (RIPA) and call to the iteration distribution procedure. The RIPA lists for each iteration, the number of distinct data references associated with each processor. It is constructed from the list of distinct distributed array access patterns encountered in each loop iteration, which have already been recorded in files *lhs_ind_file* and *rhs_ind_file* and passed as arguments to the iter_partitioner procedure as can be seen in (Fig. 6.25); the pointers to these files are stored in the array *files*. Using the RIPA the iteration distribution subroutine assigns iterations to processors.

In the next step the set of iterations assigned to a processor is split into an ordered set of of tiles; a number (*number_of_local_tiles*) of local tiles is followed by (*number_of_all_tiles* - *number_of_local_tiles* + 1) non-local tiles.

C–Processing local tiles

```
DO  tiles = 1, number_of_local_tiles
        tile_size = upbs(tile) - lwbs(1) + 1
        CALL  read_slab(files(1), edge1, lwbs(tile), tile_size)
        CALL  read_slab(files(2), edge2, lwbs(tile), tile_size)
```
C–Constructing the global reference list
```
        k = 1
        DO  i = 1, tile_size
                globref(k) = edge1(i); globref(k+1) = edge2(i); k = k + 2
        END DO
```
C–Index conversion
```
        CALL  ind_conv(dtrat, globref, locref, tile_size*2)
```
C–Executor 1
```
        k = 1
        DO  i = 1, tile_size
                x(locref(k)) = x(locref(k)) + y(locref(k+1))
                x(locref(k+1)) = x(locref(k+1)) + y(locref(k)); k = k + 2
        END DO
END DO
```

C–Processing non-local tiles

```
DO  tiles = number_of_local_tiles + 1 , number_of_tiles
        CALL  read_slab(files(1), edge1, lwbs(tile), tile_size)
        CALL  read_slab(files(2), edge2, lwbs(tile), tile_size)
```
C–Constructing the global reference list
```
        k = 1
        DO  i = 1, tile_size
                globref(k) = edge1(i); globref(k+1) = edge2(i); k = k + 2
        END DO
        CALL  localize(dtrat, sched, globref, locref, 2*tile_size, nnloc, local_size)
        CALL  zero_out_buffer(y, y(locseg_y_size+1), sched)
        CALL  Gather(y, y(locseg_y_size+1), sched)
```
C–Executor 2
```
        k = 1
        DO  i = 1, tile_size
                x(locref(k)) = x(locref(k)) + y(locref(k+1))
                x(locref(k+1)) = x(locref(k+1)) + y(locref(k)); k = k + 2
        END DO
        CALL  Scatter_add(x, x(locseg_x_size+1), sched)
END DO
```

Fig. 6.26. The Inspector and Executor

The tile description is returned in the arguments *lwbs* and *upbs* in the same way like it was provided by the primitive build_block_tiles.

Finally, two local files storing indirection arrays are created. Ordering the array elements in the files establishes alignment with the iterations in the tiles. Pointers to these local files are stored in the array *files*.

Inspector and Executor. On each processor, the inspector and executor are executed for local tiles first, and then for non-local tiles, as shown in Fig. 6.26. For each tile, contiguous blocks of values *edge1* and *edge2* can be read from the appropriate local files. The basic functionalities of the inspector and executor are described earlier.

In the phase that processes local tiles, the procedure ind_conv performs the global to local index conversion for array references that do not refer non-local elements. The local indices are used in the Executor 1 phase.

In the phase that processes non-local tiles, the list *globref*, its size and the distribution descriptor of the referenced array are used by the CHAOS procedure localize to determine the appropriate *schedule*, the size of the communication buffers to be allocated, and the local reference list *locref* which contains results of the global to local index conversion. The declaration of the arrays *x* and *y* in the message passing code allocate memory for the local segments holding the local data and the communication buffer storing copies of non-local data. The buffer is appended to the local segment. An element from *locref* refers either to the local segment of the array or to the buffer.

The Executor 2 is the final phase in the execution of a non-local tile; it performs communication described by schedules, and executes the actual computations for all iterations in the tile. The schedules control communication in such a way that execution of the loop using the local reference list accesses the correct data in local segments and buffers. Non-local data needed for the computations on a processor are gathered from other processors by the runtime communication procedure *Gather*. In the computation, the values stored in the array *x* are updated using the values stored in *y*. During computation, accumulations to non-local locations of array *x* are carried out in the buffer associated with array *x*. This makes it necessary to initialize the buffer corresponding to non-local references of *x*. To perform this action the function zero_out_buffer is called. After the tile's computation, data in the buffer location of array *x* is communicated to the owner processors of these data elements by the procedure Scatter_add which also accumulates the non-local values of *x*.

6.5.3 Experiments

This subsection presents the performance of several parallel versions of a program which includes the data parallel loop (a sweep over edges) taken from the unstructured 3-D Euler solver [185]. All the experiments were carried out on the Intel Paragon System. To be able to compare the performance of OOC program versions with the in-core versions using the Paragon's virtual memory system, the programs operated on big unstructured meshes. So, the paging mechanism of Paragon was activated in some experiments. The perfomance results are presented for four implementations of the kernel loop: nonpart-inc: the data arrays and indirection arrays are in-core and are distributed blockwise; nonpart-ooc: the data arrays are in-core (block distributed) and indirection arrays are OOC; part-inc: the data arrays and indirection arrays are in-core and data arrays and loop iterations were redistributed at runtime using the RCB partitioner; part-ooc: the data arrays are in-core, the indirection arrays are OOC and the data arrays and loop iterations were redistributed using the RCB partitioner.

In the following tables, the performance results are shown in seconds for different problem sizes and different number of processors and tiles (for OOC programs); 16 or 32 tiles were used on each processor. The symbol (p) appended to some results expresses that the Intel Paragon's paging mechanism was activated during that experiment. The tables demonstrate a big performance improvement when our OOC parallelization techniques were used in comparison with the unmodified computation using virtual memory. The significant benefit gained from partitioning is also illustrated on the performance improvement of the loop. Because of the very long execution time of some program versions, the loop was executed only once. In real CFD applications, where the loop is executed many times, a dramatical performance improvement can be achieved.

the whole program			
number of processors = 32; number of tiles = 16 (for out-of-core)			
nodes	250000	302550	360000	422500
edges	1747002	2114202	2516402	2953602
nonpart-inc	22.05	22.08	24.13	25.06
nonpart-ooc	38.00	39.06	39.43	43.51
part-inc	148.91 (p)	259.52 (p)	361.32 (p)	564.51 (p)
part-ooc	138.22	151.90	159.32	601.50(p)

only the loop			
number of processors = 32; number of tiles = 16 (for out-of-core)			
nodes	250000	302550	360000	422500
edges	1747002	2114202	2516402	2953602
nonpart-inc	12.99	13.01	12.96	11.90
nonpart-ooc	34.81	36.43	36.81	40.36
part-inc	35.80(p)	117.33(p)	160.14(p)	240.86(p)
part-ooc	10.47	10.66	14.19	120.78

the whole program				
number of processors = 64; number of tiles = 16 (for out-of-core)				
nodes	250000	302550	360000	422500
edges	1747002	2114202	2516402	2953602
nonpart-inc	22.09	22.80	24.78	30.08
nonpart-ooc	50.66	57.79	58.87	58.92
part-inc	201.79	246.89	321.97 (p)	461.82 (p)
part-ooc	230.81	245.16	260.35	269.22

only the loop				
number of processors = 64; number of tiles = 16 (for out-of-core)				
nodes	250000	302550	360000	422500
edges	1747002	2114202	2516402	2953602
nonpart-inc	6.25	9.05	10.98	13.09
nonpart-ooc	45.91	52.68	53.16	54.30
part-inc	3.78	4.61	37.28(p)	130.11(p)
part-ooc	12.71	10.92	12.74	12.14

the whole program				
number of processors = 32; number of tiles = 32 (for out-of-core)				
nodes	250000	302550	360000	422500
edges	1747002	2114202	2516402	2953602
nonpart-inc	22.05	22.08	24.13	25.06
nonpart-ooc	41.36	40.65	41.24	56.80
part-inc	148.91 (p)	259.52 (p)	361.32 (p)	564.51 (p)
part-ooc	228.97	245.65	245.74	670.81(p)

only the loop				
number of processors = 32; number of tiles = 32 (for out-of-core)				
nodes	250000	302550	360000	422500
edges	1747002	2114202	2516402	2953602
nonpart-inc	12.99	13.01	12.96	11.90
nonpart-ooc	38.06	38.10	38.73	54.14
part-inc	35.80(p)	117.33(p)	160.14(p)	240.86(p)
part-ooc	14.05	15.35	15.00	162.97

the whole program			
number of processors = 64; number of tiles = 32 (for out-of-core)			
nodes	250000	302550	360000	422500
edges	1747002	2114202	2516402	2953602
nonpart-inc	22.09	22.80	24.78	30.08
nonpart-ooc	81.31	94.26	100.41	92.52
part-inc	201.79	246.89	321.97 (p)	461.82 (p)
part-ooc	311.28	411.15	424.15	408.03

only the loop			
number of processors = 64; number of tiles = 32 (for out-of-core)			
nodes	250000	302550	360000	422500
edges	1747002	2114202	2516402	2953602
nonpart-inc	6.25	9.05	10.98	13.09
nonpart-ooc	76.84	89.26	95.69	87.82
part-inc	3.78	4.61	37.28(p)	130.11(p)
part-ooc	18.56	22.49	26.03	19.29

6.5.4 Generalization of Irregular Out-of-Core Problems

In the previous section we described parallelization steps for OOC irregular problems in which it was assumed that the data arrays could fit in the memory but other data structures describing indirections and interactions could not (because they are an order of magnitude or more larger than the data arrays). Though the above assumption is quite realistic for many problems, in this section we remove this restriction. We present an overview of parallelization steps and the corresponding runtime support in which both the data arrays as well as interaction/indirection arrays are Out-of-Core.

The main problem that must be addressed in this case is that the data arrays are also out-of-core. Consequently, unlike in the previous case, update of a value of a node (data element) may require reading the value from disk, updating it and then storing it back. Therefore, it is possible that each update may result in two I/Os if the data element is not in memory when the iteration(s) updating it are executed. For example, in Fig. 6.19, when executing iteration 10 (marked as edge 10 in the figure), if node 1 is not in memory, then it requires a read-update-write operation, resulting in two I/Os. Also, note that, execution of an iteration in the original loop (Fig. 6.18) requires updating both nodes connected to the edge representing the iteration (e.g., nodes 1 and 7 for edge 10 in Fig. 6.19). It is important to note that iterations are organized according to the edges, while node values are updated.

We propose that in order to minimize I/O, it is important to update the values of a node (data element) for the maximum number of times before it is written back, thereby minimizing number of I/Os per data element.

Thus the loop iterations must be split and reorganized to achieve the above goal. Minimizing I/O for updates does not necessarily minimize overall I/O (although it reduces it to a very large extent), because updates to nodes require reading values of the nodes connected on the other side of the edge. Therefore, scheduling tiles in such a way as to maximize reuse of the read only data, and choosing an appropriate replacement policy for the read-only tiles are also important. However, the total number of I/Os for update phase would still be much larger if the data is not reorganized and the loop is not transformed.

In the following subsection, we describe steps for parallelization of OOC irregular problems when both the data elements as well as all indirection/interaction arrays are out–of–core. We will present these steps as extensions of the steps described in the previous section.

Our approach is based on the loop transformation called the *iteration splitting loop transformation* which is depicted in Fig. 6.27. In the original loop, both nodes connected by an edge are updated by statements *S1* and *S2* in one loop iteration. In the transformed loop, in the first phase, a node with the lower index is updated only; the opposite node is updated in the second phase. The ordering of updates is achieved by an appropriate renumbering of *edge1* and *edge2* arrays, which is realized by functions *Renumber1*, ..., *Renumber4*.

If applied to the mesh depicted in Fig. 6.19, the sequence of node accesses is illustrated in Fig. 6.28.

Parallelization Steps and Runtime Support

For parallelization, steps up to step E are the same as in the previous problem described in subsection 6.5.2. At the end of step E, each processor has local files that are created using the output of a partitioner. Furthermore, iterations (indirection arrays) are distributed to minimize interprocessor communication. The only difference is that the data elements also reside in local files. We describe the proposed support for the subsequent steps below.

F. In this step the loop iterations are first split. This essentially requires partitioning the indirection arrays (or edges) into two sets. Each of these sets is then sorted according to the node number. In other words, each iteration is enumerated and sorted (thereby renumbering the iterations) so that all iterations accessing the lowest number element is grouped first followed by the second lowest and so on. That is, they are organized in the ascending order. This is illustrated in Fig. 6.29, where all iterations accessing node 1 are bunched together. Note that each processor performs this algorithm in parallel on its local iteration sets and data. The rationale behind this was alluded to earlier, where we stated that by this reorganization read-modify-write I/O operations will be minimized by

The irregular data-parallel loop *L2* in the following code

```
        PARAMETER :: NNODE = 7, NEDGE = 12, NSTEP = ...
        REAL x(NNODE), y(NNODE)              ! data arrays
        INTEGER edge1(NEDGE), edge2(NEDGE)   ! indirection arrays
L1:     DO j = 1, NSTEP                      ! outer loop
L2:         DO i = 1, NEDGE                  ! inner loop
S1:             x(edge1(i)) = x(edge1(i)) + y(edge2(i))
S2:             x(edge2(i)) = x(edge2(i)) + y(edge1(i))
            END DO
        END DO
```

is transformed into two loops as follows

```
    temp1 = Renumber1(edge1,edge2); temp2 = Renumber2(edge1,edge2);
    edge1 = temp1; edge2 = temp2
L1: DO j = 1, NSTEP
       DO iphase = 1,2
          DO i=1, NEDGE
S:          x(edge1(i)) = x(edge1(i)) + y(edge2(i))
          END DO
          IF (iphase = 1) THEN
             temp1 = Renumber3(edge1,edge2); temp2 = Renumber4(edge1,edge2);
             edge1 = temp1; edge2 = temp2
          END IF
       END DO
    END DO
```

where *temp* is a temporary array

Fig. 6.27. Iteration Splitting Loop Transformation

performing a large number of updates before requiring more I/O operations. All nodes requiring interprocessor communication are also included in this set. This allows communication to be performed only in one phase. The second set minus the iterations requiring interprocessor communication are also reorganized by sorting in the ascending order to minimize I/O.

Fig. 6.29 illustrates the iterations executed in the first phase. The edge numbers describe the iteration numbers. Iterations marked with a ' are executed on processor 2 whereas all others are executed on processor 1. For example, 7/4' means that update of node 5 is performed on processor 1 in this phase (in the executor step), whereas the update of node 3 is performed in iteration 4 of processor 2. Fig. 6.30 describes the iterations executed in the second phase. Note that no interprocessor communication is required since it is performed entirely in the first phase.

G. Once the above reorganization is done, the inspector can be executed. The inspector requires additional steps than in the previous case. Inspector

Node Accesses in the Original Loop			PHASE 1 (iphase = 1)		
i	edge1(i)	edge2(i)	i	edge1(i)	edge2(i)
1	1	5	1	1	4
2	4	5	2	1	5
3	2	3	3	1	7
4	1	4	4	2	3
5	3	6	5	2	6
6	2	6	6	3	5
7	4	6	7	3	6
8	3	5	8	3	7
9	5	6	9	4	5
10	1	7	10	4	6
11	3	7	11	5	6
12	5	7	12	5	7

PHASE 2 (iphase = 2)

i	edge1(i)	edge2(i)
1	3	2
2	4	1
3	5	1
4	5	3
5	5	4
6	6	2
7	6	3
8	6	4
9	6	5
10	7	1
11	7	3
12	7	5

Fig. 6.28. Node Accesses in the Original and Transformed Loops

essentially develops a communication schedule for the first phase (again this can be performed either in its entirety or once for each slab/tile). Furthermore, inspector for each slab is used to determine the read-only slabs schedule for each tile being updated. This information can be effectively used for prefetching read-only slabs and determining the replacement policy. The number of read-only slabs depends on the amount of memory available.

Once the schedules are developed for communication and I/O, executor may be executed.

THE FIRST PHASE

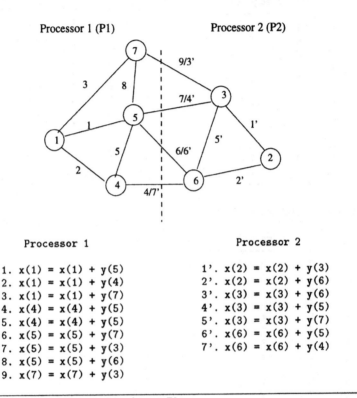

Processor 1 (P1)   Processor 2 (P2)

Processor 1

1. x(1) = x(1) + y(5)
2. x(1) = x(1) + y(4)
3. x(1) = x(1) + y(7)
4. x(4) = x(4) + y(5)
5. x(4) = x(4) + y(5)
6. x(5) = x(5) + y(7)
7. x(5) = x(5) + y(3)
8. x(5) = x(5) + y(6)
9. x(7) = x(7) + y(3)

Processor 2

1'. x(2) = x(2) + y(3)
2'. x(2) = x(2) + y(6)
3'. x(3) = x(3) + y(6)
4'. x(3) = x(3) + y(5)
5'. x(3) = x(3) + y(7)
6'. x(6) = x(6) + y(5)
7'. x(6) = x(6) + y(4)

Fig. 6.29. Iterations Executed in the 1st Phase.

H. Step H represents the executor. In this phase, for each time step, the execution is performed in two phases (once for each loop obtained after splitting).

Implementation Outline

For the class of applications considered, data and iteration partitioners must be out-of-core. They are based on so called *disk oriented partitioning algorithms*; the partitioner will have available only a limited main memory area.

Application of the parallelization strategy which was discussed in the previous paragraph is illustrated in Fig.s 6.31 and 6.33. Below, we outline the effect of individual processing steps, if they are applied to the mesh depicted in Fig. 6.19.

THE SECOND PHASE

Processor 1 (P1)   Processor 2 (P2)

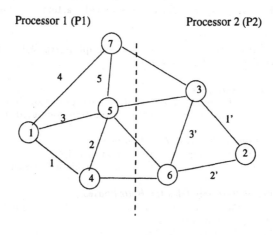

Processor 1

1. x(4) = x(4) + y(1)
2. x(5) = x(5) + y(4)
3. x(5) = x(5) + y(1)
4. x(7) = x(7) + y(1)
5. x(7) = x(7) + y(5)

Processor 2

1'. x(3) = x(3) + y(2)
2'. x(6) = x(6) + y(2)
3'. x(6) = x(6) + y(3)

Fig. 6.30. Iterations Executed in the 2nd Phase

1. *Initialization*

OOC data arrays x and y are stored in the global files that are denoted by *x_file* and *y_file*, respectively. The mesh description is stored in the file denoted by *mesh_file*; in this file, the *edge1* elements precede the *edge2* elements, as expressed below

mesh_file : {1,5,2,1,5,2,6,5,6,1,7,5, | 5,4,3,4,2,6,4,3,5,7,3,7}

2. *Data Partitioning*

A disk-oriented partitioning algorithm (referred also to as the OOC partitioner) decomposes x and y data sets to a number of partitions and allocates each partition to a processor. The produced partitions need to have the minimum number of connections to other partitions (minimum communication and remote disk access requirements) and they should have the approximately equal number of data elements (good load balancing). The distribution produced by a partitioner results in a table called the *translation table* that stores a processor reference and the logical offset

1. *Initialization*
   open (y_file, filename = 'y.dat', status = 'old', , action = 'read')
   open (mesh_file, filename = 'edge1_edge2.dat', status = 'old', action = 'read')
   open (x_file, filename = 'x.dat', status = 'old', action = 'readwrite')

2. *Disk-Oriented Data Partitioning*
   Construct out-of-core data distribution descriptor (translation table)
   which is stored in the file trans_file.

3. *Creation of local data files*
   y_file_loc and x_file_loc are constructed for each processor from y_file
   and x_file according to the new data distribution computed in step 2.

4. *Basic Disk-Oriented Iteration Partitioning*
   Construct out-of-core iteration distribution descriptor
   (iteration table) which is stored in the file iter_file_loc at each processor.

5. *Construction of Iteration Tiles for Both Phases*
   Phase1: edge1_file_ph1loc, edge2_file_ph1loc; tiles that do not need
   communication precede other tiles
   Phase2: edge1_file_ph2loc, edge2_file_ph2loc

Fig. 6.31. OOC Data and Iteration Partitioning and Reorganization of the Files

in that processor's memory for each array element. The translation table
is distributed blockwise across the corresponding local files. If we assume
that the partitioner produces the distribution depicted in Fig. 6.20, and
the translation table is distributed across two local files, each denoted by
transtab_file_loc, the local files will have the following contents:
transtab_file_loc at processor 1 (describes elements with indices 1-4) :
$$\{(1,0),(2,0),(2,1),(1,1)\}$$
transtab_file_loc at processor 2 (describes elements with indices 5-7) :
$$\{(1,2),(2,2),(1,3)\}$$

For example, information about the owner processor and offset for the
array element $y(5)$ is stored as the first record of the file *transtab_file_loc*
at processor 2; $y(5)$ has got offset 2 in the data segment of y assigned to
processor 1.

3. *Creation of local data files*
   To complete the data partitioning process, all files carrying data arrays
   are redistributed to the new data distribution described by the translation
   table. In our example, the appropriate local files are constructed from the
   files storing arrays x and y as illustrated in Fig. 6.32.
   The contents of the local files are as follows:

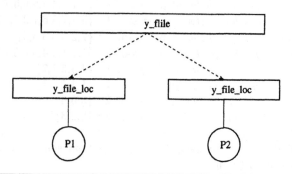

Fig. 6.32. Redistribution of Datafiles

x_file_loc at processor 1 : {x(1),x(4),x(5),x(7)}
x_file_loc at processor 2: {x(2),x(3),x(6)}
y_file_loc at processor 1 : {y(1),y(4),y(5),y(7)}
y_file_loc at processor 2 : {y(2),y(3),y(6)}

4. *Basic Disk-Oriented Iteration Partitioning*
   Loop iterations are redistributed according to the new data distribution.
   The table describing the iteration distribution is distributed blockwise
   across specific local files. In our case, two local files, *itertab_file_loc* at
   processor 1 and *itertab_file_loc* at processor 2, are created. Their contents
   are as follows (compare with Fig. 6.20):

   itertab_file_loc at processor 1 (iterations 1-6) : {1,1,2,1,2,2}
   itertab_file_loc at processor 2 (iterations 7-12) : {2,2,1,1,2,1}

   For example, the iteration 9 is currently assigned to processor 1 (this
   information is stored as the third record of the file *itertab_file_loc* at processor 2).

5. *Construction of Iteration Tiles for Both Phases*
   Iterations assigned to individual processors by the basic partitioning algorithm in the previous step are reorganized and formed into tiles; on
   each processor one set of tiles is formed for Phase 1 and another one
   for Phase 2. Tiles that do not need communication (no communication
   buffers have to be allocated) may be larger and they precede tiles that
   include inter-processor communication. The resulting sets of tiles determine the distribution of the edge data across local files. Below, local edge
   files are enumerated as they are needed in the corresponding phases, and
   their contents are described:
   – Phase1: *edge1_file_ph1loc, edge2_file_ph1loc*
   – Phase2: *edge1_file_ph2loc, edge2_file_ph2loc*

       edge1_file_ph1loc at processor 1 : {1,1,1,4,4,5,5,5,7}
       edge2_file_ph1loc at processor 1 : {5,4,7,5,6,7,3,6,3}
       edge1_file_ph1loc at processor 2 : {2,2,3,3,3,6,6}
       edge2_file_ph1loc at processor 2 : {1,1,1,4,4,5,5,5,7}
       edge1_file_ph2loc at processor 1 : {4,5,5,7,7}
       edge2_file_ph2loc at processor 1 : {1,4,1,1,5}
       edge1_file_ph2loc at processor 2 : {3,6,6}
       edge2_file_ph2loc at processor 2 : {2,2,3}

Note that accesses to the elements of y exhibit some sort of (irregular) temporal data locality.

With reorganization of the files, the data is prefetched as close to the processors that need them; moreover, they can be read in blocks and the application of the data sieving technique [78] may be more efficient.

6. *Inspector/Executor - Phase 1*

Information describing the tiles which are going to be applied in Phase 1 and Phase 2 is stored in the descriptors *tile_ph1_desc* and *tile_ph2_desc*, respectively. The slab of *edge1* read at line *F2* is analyzed by procedure *analyze* (line *F3*) which determines the slab (tile) of x (described by the lower bound *x_lwb* and size *x_size*) to be read from file *x_file_x* at line *F4*. Moreover, procedure *analyze* precomputes the local indices and stores the results of this index conversion in array *locref_x*. The slab of *edge2* read at line *F5* is used by the procedure *disk_mem_localize* (line *F6*) to compute two data structures, *disk_sched* and *mem_sched*, that describe the required I/O and communication for y; the numbers of elements to be transferred are stored in variables *n_disk* and *n_mem* and the results of the index conversion in array *locref_y*. The elements of y are gathered from disk and other processors by procedures *disk_gather* and *mem_gather*. The incoming data elements are stored in the memory portion allocated for y; the first part of the portion is filled by the elements from disk and the second one by the elements from other processors. The executor is represented by the statements at lines *F9 – F11*. The values of x computed by the inspector are stored in the local files *x_file_loc*.

7. *Inspector/Executor - Phase 2*

In Phase 2, no inter-processor communication is needed. Therefore, only the disk schedule is computed (line *S6*) that passed to the procedure *disk_gather* (line *S7*) that gathers the y elements from disk.

6. Out-of-Core Inspector/Executor - Phase 1

```
F1:     DO tile = 1, number_of_tiles_phase1
F2:         CALL read_slab(edge1_file_ph1loc, edge1, tile_ph1_desc, tile)
F3:         CALL analyze(edge1, locref_x, x_lwb, x_size)
F4:         CALL read_slab(x_file_loc, x, x_lwb, x_size)
F5:         CALL read_slab(edge2_file_ph1loc, edge2, tile_ph1_desc, tile)
F6:         CALL disk_mem_localize(disk_sched, mem_sched, locref_y, edge2, &
                                   n_disk, n_mem)
F7:         CALL disk_gather(disk_sched, y_file, y(1))
F8:         CALL mem_gather(mem_sched, y(n_disk+1))
F9:         DO i = 1, tile_ph1_desc%tile_size(tile)
F10:            x(locref_x(i)) = x(locref_x(i)) + y(locref_y(i))
F11:        END DO
F12:        CALL write_slab(x_file_loc, x, x_lwb, x_size)
F13:    END DO
```

7. Out-of-Core Inspector/Executor - Phase 2

```
S1:     DO tile = 1, number_of_tiles_phase2
S2:         CALL read_slab(edge1_file_ph2loc, edge1, tile_ph2_desc, tile)
S3:         CALL analyze(edge1, locref_x, x_lwb, x_size)
S4:         CALL read_slab(x_file_loc, x, x_lwb, x_size)
S5:         CALL read_slab(edge2_file_ph2loc, edge2, tile_ph2_desc, tile)
S6:         CALL disk_localize(disk_sched, locref_y, edge2, n_disk)
S7:         CALL disk_gather(disk_sched, y_file, y(1))
S8:         DO i = 1, tile_ph2_desc%tile_size(tile)
S9:             x(locref_x(i)) = x(locref_x(i)) + y(locref_y(i))
S10:        END DO
S11:        CALL write_slab(x_file_loc, x, x_lwb, x_size)
S12:    END DO
```

Fig. 6.33. Out-of-Core Inspector/Executor

6.6 Support for I/O Oriented Software-Controlled Prefetching

In general, *prefetching* ([194]) is a technique for tolerating memory latency by explicitly executing prefetch operations to move data close to the processor before it is actually needed. It presents two major challenges. First, some sophistication is required on the part either the programmer, runtime system, or (preferably) the compiler to insert prefetches into the code. Secondly, care must be taken that the overheads of prefetching, which include additional instructions and increased memory queuing delays, do not outweigh the benefits.

In prefetching discussed in the context of superscalar or shared-memory architectures, a block of data is brought into the cache before it is actually referenced. The compiler tries to identify data blocks needed in the future and, using special instructions, tells the memory hierarchy to move the blocks into the cache. When the block is actually referenced it is found in the cache, rather than causing a cache miss.

On the other hand, efficiency of the parallel I/O operations can be dramatically increased if the compiler and runtime system are able to prefetch data to disks from which the transfer to the target processors is most optimal. Therefore, in this section, we introduce *I/O extended prefetching* helps to match disk, memory and processor speeds in parallel systems. The aim of this kind of prefetching is to reduce memory latency by explicit (i.e., runtime system driven) execution of disk and/or network prefetch operations in order to move data (e.g., file blocks, OOC array slabs, startup/restart volumes) as close as possible to the processor which is going to need the data.

Two major problems have to be tackled in this context. First, some sophistication is required by either the programmer, runtime system, or the compiler ([127]) to insert appropriate prefetch requests into the code. An interesting problem in this context is to avoid prefetching in situations where the data item could be invalidated, e.g., by a modification done by another processing element in the time interval between the prefetch operation and the actual use of the prefetched data item. Secondly, care must be taken that the overheads for prefetching, which include additional instructions and increased memory queuing delays, do not outweigh the benefits. Consequently, the prefetching subsystem has to employ some kind of cost/benefit estimation on which the actual prefetching process has to rely. On most occasions, this cost/benefit estimation is based on a formal machine model (see [204]) and some heuristics used to cut down intractably large optimization search spaces.

In this section, we deal with two methods to support the I/O extended prefetching. The first method is based on the compile time program analysis which decides about placement of data prefetching directives. In the second method, the compiler passes to the runtime system information about accesses to OOC structures and predicted execution time in form of an annotated program graph. The runtime system uses this information for derivation of his prefetching strategy.

6.6.1 Compiler-Directed Data Prefetching

The compiler can determine the earliest point in the execution of a program at which a block of data can be prefetched for a given processor. This determination is based on the data dependence and control dependence analysis in the program.

The analysis methods proposed by Gornish ([127, 128, 129]) for optimizations of programs developed for shared-memory systems can be applied to out-of-core programs. Here, we informally outline the basic ideas.

Pulling Out a Reference from a Loop. The example below gives an illustration of a prefetch statement.

---

the input form of the loop; OOC arrays A and B are BLOCK *distributed*

```
DO  i=2, n-1
    A(i) = (B(i-1) + B(i+1)/2
END DO
```

the transformed form of the loop

```
Prefetch  (B($lA(p)-1 : $uA(p)+1))
...  OOC version of the above do loop ...
```

If the owner computes rule is applied, processor p will execute iterations $i \in (\$lA(p), \$uA(p)))$, where $\$lA(p)$ and $\$uA(p))$ denote the upper and lower bounds of the local segment of A on p, respectively. The references to $B(i-1)$ and $B(i+1)$ on processor p uses elements with indices $\$lA(p) - 1$ through $\$uA(p) + 1$. Therefore, we can prefetch these elements into the buffer or onto the local disk before the execution of the slabs starts.

For example, we can prefetch a reference used in a conditional statement, such as in the example below:

```
DO  i=1,n
      IF  (A(i) > 5)  THEN
         ...
      END IF
END DO
```

We say that we *pull out* a reference from a loop, if all the elements, generated by the reference in the loop, can be prefetched before the start of the loop.

Prefetch Suppression. We say that a prefetch is suppressed if it cannot be pulled further back in a program without the possibility of either:

1. prefetching data that might not be used
2. some of the prefetched data being modified before it is used.

Restriction 1 is generally caused by control dependences. Restriction 2 is caused by data dependences.

Control Dependence. We do not pull out a reference whose execution is controlled by a conditional, since we might be prefetching unnecessary data or, worse, non-existent data. Consider the example below.

```
DO i=1,10
   IF (i<5) THEN
        A(i) = B(i) + 1
   END IF
END DO
```

If we were to pull out the reference to $B(i)$ from the loop, there would be two potential problems:

1. We would prefetch elements 1 through 10 of array B, and would, therefore, be prefetching five data that would not be used. This is an important concern, since we would be doing unnecessary work. If we were to prefetch too much data that would not be used, we might offset the gains derived from prefetching.
2. A more serious problem might exist. If B were declared to contain only elements 1 through 5, we would be prefetching data that did not exist and would encounter an error.

These two problems cause prefetches to be suppressed by control dependence. Or a heuristic method may be used if the second problem does not occur.

Data Dependence. Data dependencies can cause data to be prefetched unnecessarily or, even worse, prefetched data to be invalid when it is used. The example below shows how this might happen.

```
      DO i = 1, n
S1:      A(i) = B(i) + 1
S2:      C(i) = A(i) + 1
      END DO
```

If we pulled out the reference to $A(i)$, in statement $S2$, from the loop, the prefetched data would be modified before it would be used.

6.6.2 Data Prefetching Based on Prefetch Adaptivity

Prefetch adaptivity involves adapting when prefetches are issued for different data, based on program and system characteristics. Prefetch adaptivity includes both compile time and runtime features. At compile time, prefetches are scheduled, such that the prefetched data should be available when needed. At runtime, the issuing of prefetches is dynamically adapted, based on system characteristics such as disk access latency. Both runtime characteristics and the variability between different programs make prefetch adaptivity very important for an effective data prefetching algorithm.

In our approach, the compile-time knowledge about loops and other out-of-core code parts is represented by a graph structure called I/O Requirement Graph (IORG) containing information to be used to deduce hints at runtime (see for example [210] or [131] on hints for I/O prefetching and caching in general). This graph is constructed incrementally during the compilation process and written to a file at its end. The IORG is used by VIPIOS to direct the prefetching and data reorganization processes.

```
START:                              START:
    SSQ1                                CALL VIPIOS_branch('b1')
S1: A(lhs1) = ... B(rhs1) ...           SPMD_SSQ1
    SSQ2                                OOC_S1
IF: IF (condition) THEN                 SPMD_SSQ2
    SSQ3                            IF (condition) THEN
S2:    B(lhs2) = ... B(rhs2) ...        CALL VIPIOS_branch('b2')
S3:    C(lhs3) = ... B(rhs3) ...        SPMD_SSQ3
    ELSE                                OOC_S2
S4:    D(lhs5) = ... B(rhs5) ...        OOC_S3
       SSQ4                         ELSE
S5:    E(lhs6) = ... B(rhs6) ...        CALL VIPIOS_branch('b3')
              ... A(rhs7) ...           OOC_S4
    END IF                              SPMD_SSQ4
    SSQ5                                OOC_S5
END:                                END IF
                                    CALL VIPIOS_branch('b4')
                                    SPMD_SSQ5
                                END:
        (a) Source program              (b) SPMD program
```

Fig. 6.34. IORG Base Information

We denote the IORG by the triple $G=(N,E,s)$ where (N,E) is a directed graph, $s \in N$ is the start node with indegree 0, and each $n \in N$ is in the transitive cover s^*. The set of nodes N consists of nodes represented in-core program parts and out-of-core ones. In the IORG, code segments are condensed into single nodes encompassing the following components:

– Branch label.
  In general, IORG structures contain patterns with multiple branches including symbolic information to be resolved at runtime. A unique label is assigned to each branch which is included in all nodes belonging to the

branch. The VIPIOS is informed at runtime about the branch selection
by a compiler-inserted call from the SPMD program. In the IORG these
call points are annotated. When the executing program goes through the
corresponding instruction sequence, the call is issued. Runtime data for
the substitution of symbolic information is transferred in a similar way.
Synchronization points are determined in the context of data flow analysis
thus finding process states in which values can be bound to these symbols
in the IORG.

- Time cost estimation.
  It estimates how much processor time is used by a program fragment represented by this node.
- Descriptions of array sections handled by a node. I/O array sections are read from or written to the parallel array database contained in the VIPIOS repository.

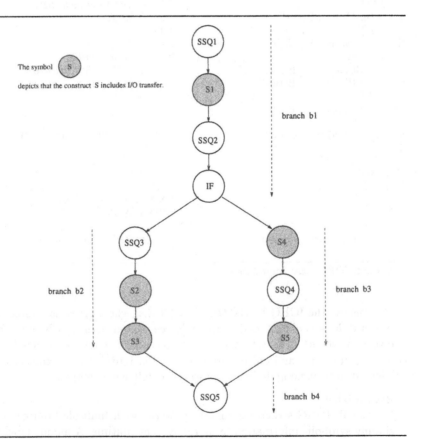

Fig. 6.35. Basic IORG Graph

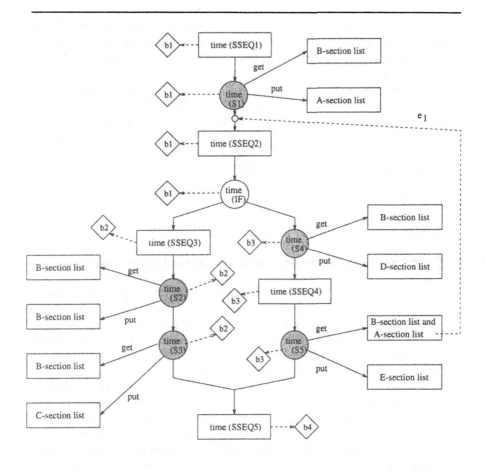

Fig. 6.36. Expanded IORG Graph

The construction of an IORG is sketched in the following example. Figure 6.34.(a) introduces a fragment of the input out-of-core code, where SSQ1–SSQ5 denote in-core code parts (statement sequences) and S1–S5 are array statements that include out-of-core arrays A–E. Pseudocode of the appropriate SPMD program is introduced in Figure 6.34.(b), where SPMD_SSQi denotes an SPMD form of SSQi and OOC_Si the out-of-core form of Si. In this example, the IORG consists of four branches (denoted by b1, ..., b4). VIPIOS is informed about the actual selection of a particular branch by the *VIPIOS_branch* procedure call issued in the SPMD program. The corresponding IORG graph is depicted in Figure 6.35 and its expanded form in Figure 6.36. Edge e_l points to the point of the program that constrains the movement of the prefetch for A further back. With the node *S1*, for example, the following information is associated: the program part represented by this node is in branch *b1* and reads a sequence of sections of array B that are denoted by *B-list* and writes a sequence of sections of array A that are denoted by *A-list*. The corresponding execution time estimates are denoted by *time(S1)*.

Prefetching cost estimation is based on interconnection network distance and workload as well as I/O channel workload. Prefetching benefit estimation includes information about the probability of an actual use of prefetched items. In a control flow sequence, this probability equals 1, whereas in a control flow branch structure, the probability is less then 1. This yields to a concept commonly known as speculative (prefetching) work.

BIBLIOGRAPHICAL NOTES

Compiling out-of-core data-parallel programs is a fairly new topic and there has been little research in that area. A compiler supporting out-of-core data-parallel programs is described in [206]. They are developing a compiler, extension of the Fortran D compiler, that inserts explicit statements to move array data in and out of memory, guided by I/O distribution statements similar to Fortran D's in-memory data distribution statements. They report on early experiments (partially compiled by hand) which show superior performance compared to a program using virtual memory.

Research in compiler support for out-of-core HPF programs that uses the PASSION runtime library [80] is reported in [40, 244]. They propose two new directives to specify which arrays are to be out-of-core and how much memory is available in a processor to hold out-of-core array data. Ideas from [244] and [206] are combined in [43] which describes a runtime system employing the two-phase I/O approach for handling array I/O. Cormen and Colvin are developing a compiler-like preprocessor for out-of-core C*, called ViC* [88], which translates out-of-core C* programs to standard C* programs with calls to a runtime library for I/O.

7. Runtime System for Parallel Input-Output

In this chapter, we investigate the I/O problem from a runtime support perspective. We focus on the design of a novel parallel I/O support, called VIPIOS (VIenna Parallel I/O System). VIPIOS is coupled to the compiler modules that process I/O operations and handle out-of-core (OOC) structures. Specifically, our design is presented in the context of the Vienna Fortran Compilation System (VFCS).

After a short description of the architecture of the VFCS and VFCS-VIPIOS interconnection, we introduce a formal model, which forms a theoretical framework on which the VIPIOS design is based. Then we describe the VIPIOS design which is the core of this chapter. The VIPIOS design is partly influenced by the concepts of parallel database technology.

7.1 Coupling the Runtime System to the VFCS

The VFCS overall architecture including VIPIOS is depicted in Fig. 7.1. This figure also shows how the information is passed from Vienna Fortran programs through VFCS to VIPIOS.

Vienna Fortran programs are processed by the frontend and transformed into an internal representation. All parallelization transformations are performed on the internal representation.

VFCS is an interactive system. During the program transformation process, the user is able to inspect the internal information, supply special information to the system and select transformations. The ability of an interactive system to provide information about the program on a selected level is useful during the parallelization process as well as for the development of new transformations. For example, it is easy to find inefficiencies via the inspection of the computed overlap information, or to identify parallelization inhibiting factors by analyzing data dependences.

The kernel implements the user interface. The user may activate other system parts and select analysis services and transformations. He is able to inspect the internal information via the services of the information component. Since the program database contains only the actual version of the program, VFCS provides the additional service to save program versions.

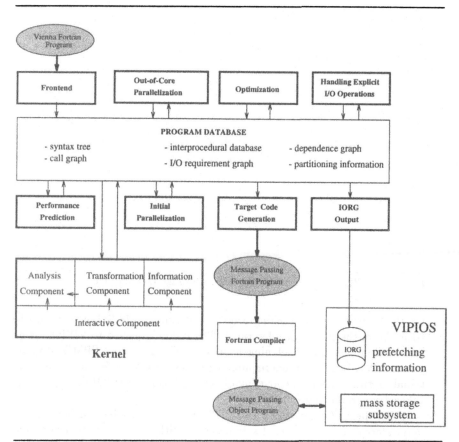

Fig. 7.1. VIPIOS in the VFCS Environment

The user may thus return to an earlier state of the parallelization process if the performed transformations were not successful.

VFCS is built around an interprocedural database. The database supports the effective development and parallelization of modular programs by providing a knowledge base containing information about application programs and their (independently compiled) units and modules. Information about objects may include control flow graphs, results of data flow analysis, the call graph, dependence graphs, and results of data distribution analysis. Furthermore, the database defines a common interface for other tools supporting VFCS. One of them is the *performance prediction tool* [108] which statically computes a set of optional parameters that characterize the behavior of the parallel program.

Program analysis and performance prediction phases extend the internal representation with information used for optimizing transformations and construction of an I/O requirement graph (IORG) which is used for guid-

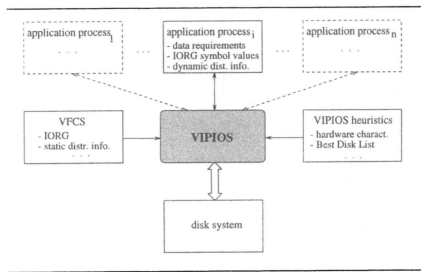

Fig. 7.2. VIPIOS Viewed as Abstract Mass Storage I/O Machine

ing the prefetching and caching processes at runtime. In Subsection 6.6, the structure of the IORG was described. The IORG is passed to VIPIOS as an information base for prefetching heuristics. The generated SPMD program passes to VIPIOS the actual requirements on section transfer as well as additional runtime data replacing symbolic information in the IORG (e.g., actual values for variable names).

From the VFCS point of view, VIPIOS can be considered as an *abstract mass storage I/O machine*. Its interaction with the compilation and runtime environment is depicted in Fig. 7.2 where the term application process denotes a running SPMD program.

7.2 Data Mapping Model

The model introduced in this section describes the mapping of the problem specific data space starting from the application program data structures down to the physical layout on disk across several intermediate representation levels. Here, we repeat several definitions from Section 3.1 to provide a more compact presentation of the model.

Parallelization of a program Q is guided by a mapping of the *data space* of Q to the processors and disks. The data space $\mathcal{A}$ is the set of declared arrays of Q.

Definition 7.2.1. Index Domain

Let $A \in \mathcal{A}$ denote an arbitrary array. The index domain *of A is denoted by $\mathbf{I}^A$. The* extent *of A, i.e. the number of array elements is denoted by ext^A.*

The index domain of an array is usually determined through its declaration.

In a lot of languages, only array elements or whole arrays may be referenced. On the other hand, some languages like Fortran 90 provide features for computing subarrays. A subarray can be either characterized as an array element or an array section.

Definition 7.2.2. Index Domain of an Array Section

Let us consider an array A of rank n with index domain $\mathbf{I}^A = [l_1 : u_1, \ldots, l_n : u_n]$ where the form $[l_i : u_i]$ denotes the sequence of numbers $(l_i, l_i + 1, \ldots, u_i)$. The index domain of an array section $A' = A(ss_1, \ldots, ss_n)$ is of rank n', $1 \leq n' \leq n$, iff n' subscripts ss_i are section subscripts (subscript triplet or vector subscript) and the remaining subscripts are scalars. For example the index domain of the Fortran 90 array section $A(11 : 100 : 2, 3)$ is given by $[1 : 45]$.

The extended index domain of the array section A', denoted by $\bar{\mathbf{I}}^{A'}$, is given by $\bar{\mathbf{I}}^{A'} = [ss_1', \ldots, ss_n']$ where $ss_i' = ss_i$, if ss_i is a section subscript, and $ss_i' = [c_i : c_i]$ if ss_i is a scalar subscript that is denoted by c_i. For example the extended index domain of the array section $A(11:100:2,3)$ is given by $[11:100:2,3:3]$.

7.2.1 Mapping Data Arrays to Computing Processors

Data distribution is modeled as a mapping of array elements to (non-empty) subsets of processors. The notational conventions introduced for arrays above can also be used for *processor arrays*, i.e., $\mathbf{I}^H$ denotes the index domain of a processor array H.

A *distribution* of an array maps each array element to one or more processors which become the *owners* of the element, and, in this capacity, store the element in their local memory. Distributions are modeled by functions between the associated index domains.

Definition 7.2.3. Distributions

Let $A \in \mathcal{A}$, and assume that H is a processor array. A distribution of the array A with respect to H is defined by the the the mapping:

$$\mu_H^A : \mathbf{I}^A \to \mathcal{P}(\mathbf{I}^H) - \{\phi\}, \quad \text{where } \mathcal{P}(\mathbf{I}^H) \text{ denotes the power set of } \mathbf{I}^H$$

7.2.2 Logical Storage Model

Definition 7.2.4. File

A file F_L consists of a sequence of logically connected information units, which we call records r.

Definition 7.2.5. Record

A record r consists of a fixed number m of typed data values (attributes).

$$r = (A_1, A_2, ...A_m)$$

One record can store, for example, one float number or a complete data structure.

Definition 7.2.6. File View

The sequence of data elements which results from a write operation of an array A to a file F_L and is assumed by subsequent read operations is described by the function:

$$\alpha^A \; : \; \mathbf{I}^A \to [1 : ext^A], \; (e.g., \; the \; standard \; Fortran \; record \; ordering \; may \; be \; applied)$$

7.2.3 Physical Storage Model

Definition 7.2.7. Disk

A disk d is the physical storage medium of interest. It is organized into storage units of fixed size (buckets). Let

$$D = \{d_1, ..., d_N\}$$

denote the set of all available disks.

Definition 7.2.8. Bucket

A bucket is a physical, contiguous sequence of bytes of a fixed size. It is the smallest unit, which the operating system of the underlying machine can transfer at once between the external storage medium and the main memory of a processor. Let B_j denote the set of all buckets of a disk d_j. The set B of all buckets available for storing users' data is defined by the union of all buckets B_j of all disks. Therefore,

$$B = \bigcup B_j, \; where \; 1 \le j \le |D| \; \; and \; \; |D| \; denotes \; the \; number \; of \; disks$$

Definition 7.2.9. Physical file

A physical file F_P is defined by a fixed number of physical blocks or buckets, which are stored on the external storage medium.

Definition 7.2.10. Declustering function

A declustering function δ defines the record distribution scheme for D, according to one or more attributes and/or any other distribution criteria,

$$\delta : R \to D, \; where \; R \; is \; the \; set \; of \; records \; and \; D \; the \; set \; of \; disks.$$

In other words, the declustering function δ partitions the set of records R into disjoint subsets R_1, R_2, ... $R_{|D|}$, which are assigned to disks.

Definition 7.2.11. Organization function

The organization function o_d gives the buckets B_d of the disk d, where a specific record is stored.

$$o_d : R \to B_d, \quad \text{where } B_d \text{ are the buckets of disk } d$$

Each disk has its own organization function. Therefore a family of organiza-tion functions exists, i.e. $o_1, o_2, ... o_{|D|}$.

Definition 7.2.12. Array area of a file

An array area E^A denotes the set of file records that store the elements of an array A or array section A'.

It is now possible to redefine the declustering and organization functions for array areas.

Definition 7.2.13. Areal declustering function

An areal declustering function Δ defines the set of disk drives D, which is spanned by a given area E^A. In other words, with $\mathcal{E}$ denoting the set of all possible areas of an array A

$$\Delta : \mathcal{E} \to \mathcal{P}(D) - \{\phi\}$$

Definition 7.2.14. Areal organization function

The areal organization function O_d defines the set of buckets of disk d, where the records covered by the given area E^A are stored, i.e.

$$O_d : \mathcal{E} \to \mathcal{P}(B_d) - \{\phi\}, \quad \text{where } B_d \text{ is the set of buckets of disk } d$$

Every disk has its own area organization function. Therefore a family of area organization functions exists, i.e. $O_1, O_2, ... O_{|D|}$.

Definition 7.2.15. Query mapping function

The query mapping function q gives the set of buckets B over all disks defined by a given area, i.e.

$$q : \mathcal{E} \to \mathcal{P}(B), \quad \text{where } B \text{ is the union of buckets over all disks}$$

The query mapping function can be defined by a combination of the areal declustering Δ and the areal organization function O, i.e.

$$q(E^A) = \bigcup O_i(E^A), \quad \text{where } i \in \Delta(E^A)$$

7.2.4 Example of the Model Interpretation

In Fig. 7.3 the meaning of the terms defined in the previous paragraphs is shown by a practical example of administrating 2 arrays A_1 and A_2 by 3 and 2 areas (or slabs) respectively declustered among 3 disks d_1, d_2, and d_3.

The declustering functions δ_{LF1} and δ_{LF2} determine the declustering scheme of the 2 logical files LF1 and LF2 to the physical files F1, F2, and F3 on the available disks d_1, d_2, and d_3. The mapping of the 3 resp. 2 slabs of the files LF1 and LF2 is defined by the areal declustering functions Δ_{A11}, Δ_{A12}, ..., Δ_{A22} for each slab. The physical location on the disks of the buckets

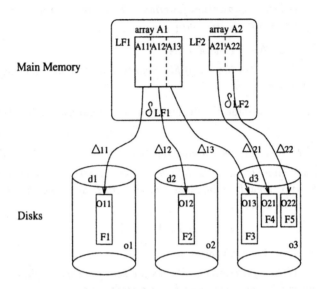

Fig. 7.3. Illustration of the Storage Model on Two Out-of-Core Arrays

containing the data sets of the 2 arrays is given by the organizational functions o_1, o_2, and o_3 for each disk (representing the data structure) and the areal organization functions $O_{11}, O_{12}, ..., O_{22}$ for the slabs (representing the query function for a given regular section of the stored array).

7.3 Design of the VIPIOS

The I/O runtime system VIPIOS introduced in this section provides support for parallel access to files for read/write operations, optimization of data-layout on disks, redistribution of data stored on disks, communication of out-of-core[1] (OOC) data and many optimizations including data prefetching from disks based on the access pattern knowledge extracted from the program by the compiler or provided by a user specification. In the first implementation, VIPIOS provides a mass storage support for VFCS.

7.3.1 Design Objectives

The goals of the VIPIOS can be summarized as follows.

---

[1] Mass storage I/O is also necessary to handle accesses to huge arrays whose parts must be kept on disks due to main memory constraints.

1. Efficiency. The objective of runtime optimizations is to minimize the disk access costs for file I/O and OOC processing. This is achieved by a suitable data layout on disks and data caching policies. Specifically, VIPIOS aims at optimizing data locality, storing data as close as possible to the processing node, and efficient software-controlled prefetching to partially hide memory latency by explicit (i.e., runtime system driven) execution of disk and/or network prefetch operations. Prefetching moves data (e.g., file blocks, OOC array sections) as close as possible to the processor which is going to need the data.
2. Parallelism. All file data and meta-data (description of files) are stored in a distributed and parallel form across multiple I/O devices. The user and the compilation system have the ability, in the form of hints, to influence the distribution of the file data.
3. Scalability. The architecture of VIPIOS is inherently scalable. Control of the whole system is decentralized.
4. Portability. VIPIOS is able to provide support for any HPF compilation system that supplies all the interface information in the required format. In order to interface a large class of computer systems, VIPIOS strongly relies on standards, like MPI [247] and MPI-IO [87].

7.3.2 Process Model

The framework VIPIOS distinguishes between two types of processes: *application processes* and *VIPIOS servers*. The process model is depicted by Fig. 7.4.

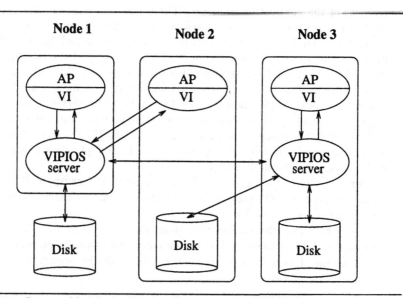

Fig. 7.4. Process Model: Application Processes and VIPIOS Servers

The application processes are created by the VFCS according to the SPMD paradigm. As far as VIPIOS is concerned, every application process is completely independent of every other application process. This independence means that VIPIOS does not impose any communication requirements on a user's application. As a result, applications may use whichever communication software (e.g., MPI, PARMACS, CHAOS) is most suitable to the given problem.

The VIPIOS servers run independently on all, a number or dedicated nodes and perform the data and other requests of the application processes. The number and the location of the VIPIOS servers are dependent on the underlying hardware architecture (disk arrays, local disks, specialized I/O nodes, etc.), the system configuration (number of available nodes, types of available nodes, etc.), the VIPIOS system administration (number of serviced nodes, disks, application processes, etc.) or the user needs (I/O characteristics, regular, irregular problems, etc.).

The VIPIOS servers are similar to data server processes in database systems. For each application process one unique VIPIOS server is assigned and accomplishes the data requests. It is also possible that one VIPIOS server services a number of application processes. In other words one-to-one or one-to-many relationships are supported.

The VIPIOS call interface *VI*, which is linked with the application process *AP*, handles the communication with the assigned VIPIOS server *VS* and other VIPIOS servers, if necessary. *VI* is implemented by the VIPIOS runtime library which runs on each compute processor. The run-time library receives I/O requests from the application, translates them into lower-level requests, and passes them (as messages) directly to the appropriate servers. The run-time library then handles the transfer of data between the VIPIOS servers and memory of the application processes. The connection between the 2 servers illustrates that both can request each other in accessing remote data.

In Fig. 7.5 different architectural models are summarized. Both local disk architectures, as disk *D2* and *D3* (nodes 2 and 3), and global disk systems as disk *D1* are supported (the surrounding frame depicts a physical processing node). Furthermore, it can be seen that varying process models are feasible. On the one hand, application process *AP3* on node 3 is serviced by VIPIOS server *VS2* running on the same node, which in turn administrates the local disk *D3* and the remote disk *D2* (local to node 2). On the other hand, the server process *VS1* handles the requests of the local application process *AP2* and accesses the global disk only.

For each application process, all data requests are transparently caught by the assigned VIPIOS process. The transfer of locally or remotely retrieved data is organized by this process.

The VFCS provides information about the problem specific data distribution, the slabs of the out-of-core data structures, and the presumed data access profile. Based on this information, the VIPIOS organizes the data

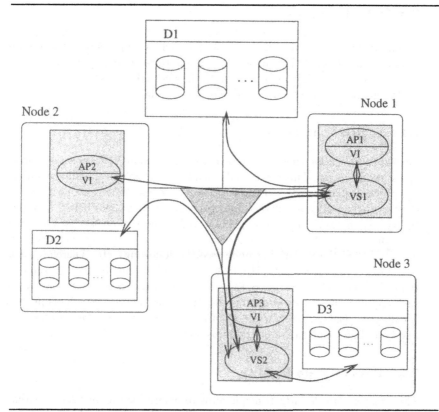

Fig. 7.5. Process Model on Top of a Disk Architecture with Local and Global Disk Systems

and tries to ensure high performance access to stored data. Additional data distribution and usage information can be provided by the Vienna Fortran programmer using new language constructs. This type of information allows the VFCS/VIPIOS system to parallelize read and write operations by selecting a well-suited data organization in the files.

An important advantage of the proposed framework is the transparence of the architecture to the application programmer as well as to the Vienna Fortran compiler developer.

Summing up, the VIPIOS, as a component of the VFCS, is responsible for the organization and maintenance of all data held on the mass storage devices.

7.4 Data Locality

The main design principle of the VIPIOS to achieve high data access performance is *data locality*. This means that the data requested by an application process should be read/written from/to the best-suited disk.

We distinguish between logical and physical data locality.

7.4.1 Logical Data Locality

Logical data locality denotes to choose the best suited VIPIOS server for an application process. This server is defined by the topological distance and/or the process characteristics. In most cases the access time is proportional to the topological distance of the application process to the VIPIOS server in the system network. It is also possible that special process characteristics can influence the VIPIOS server performance, like available memory, disk priority list of the underlying node (see the following), etc. Therefore, it is also possible that a remote VIPIOS server could provide better performance than a closer one. However, only one specific VIPIOS server is chosen for each application process, which handles the respective requests. This process is called the *buddy server*, while all other servers are called *foe servers*. The buddy and foe relation is illustrated by Fig. 7.6.

7.4.2 Physical Data Locality

The *physical data locality* principle aims to determine the disk set providing the best (mostly the fastest) data access. For each node an ordered sequence of the accessible disks of the system is defined (the *best disk list, BDL*), which orders the disks according to their access behavior. Disks with good access characteristics precede disks with bad access characteristics in this list. This is defined by technical disk characteristics, like seek time, transfer rate, etc. and/or by the topological location in the system architecture. A unique BDL is defined for each node. Thus the VIPIOS server chooses from the BDL the actual disk administrating the data of a specific application process. In most cases it will choose the disk(s) (note that it can be more than one) both according to the BDL of the node it is executing on and the physical restrictions of the disks (memory requirements, work load, etc.).

7.5 Two-Phase Data Administration Process

The data administration process of a VIPIOS server can be divided into 2 phases, the *preparation* and the *administration phase* (see Fig. 7.7).

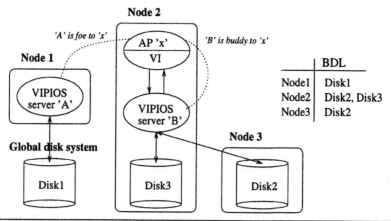

Fig. 7.6. "Buddy" and "Foe" Relation Between Application Processes and VIPIOS Servers

7.5.1 Preparation Phase

The *preparation phase* prepares the administrated data according to the distribution of the data structures stored in the files, the presumed access profile provided by the IORG graph (see Section 6.6.2) and the physical restrictions of the system (available main memory, disk space, etc.). The input data for this phase are partly prepared in the program compilation process. This phase is performed during the system startup time and precedes the execution of the application processes. In this phase, the physical data layout schemas are defined, the actual VIPIOS server process for each application process and the disks for the stored data according to the locality principles are chosen. Further, the data storage areas are prepared, the necessary main memory buffers allocated, etc.

7.5.2 Administration Phase

The *administration phase* accomplishes the I/O requests of the application processes. It is obvious that the preparation phase is the basis for good I/O performance in the administration phase. All optimizations are just performed in the administration phase. They are discussed in more detail in the following paragraphs.

7.6 Basic Execution Profile

All write and read requests are resolved by the VIPIOS. The application process requests the buddy server running on the same or another node to

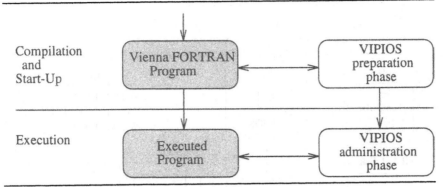

Fig. 7.7. Two-Phase Data Administration Process

perform the file operations. The data server writes the data according to the respective data distribution scheme of the logical file connected with the data to be transferred. In the simplest form this would be the local disk. Remember that the VIPIOS always aims for the data locality principle, which tries to guarantee optimal performance for the writing and reading the stored information. It is possible that the buddy server, which should perform the physical write, cannot accomplish the request due to an exception situation, e.g. a full disk system. In this case it sends the data to another server (one of the foe servers) according to an underlying exception handling scheme of the system. The information of the physical file declustering scheme is stored by the buddy server. For read requests the buddy server checks, if the requested data is stored on one of its disks. If it is not, it sends the request to the respective foe servers, which physically administrates the requested data set. This server accomplishes the request in turn, accesses the data and sends it to the application process which originally sent the request.

The procedures applied to resolve write and read requests are formally specified in the next two paragraphs using the concepts introduced in Section 7.2.

Write array section. To write a two-dimensional array section of reals, for example, every processor executes the following (simplified) code:

```
...
REAL  A(10, 1 : slabsize)
...
CALL  VIPIOS_Put_Sec(A, section_description, ... )
...
```

The procedure *VIPIOS_Put_Sec* of the VIPIOS interface (i.e. *VI* in Fig. 7.4) executes a write request[2] for the requested regular section of the array A. In most cases the regular section comprises one contiguous area. This can be depicted by

---

SUBROUTINE VIPIOS_Put_Sec(A, section_description, ...)
   contact buddy server
   ! let a be a file array area corresponding to the section of A
   which has to be written into the mass storage system
   write-request(a)

During the preparation phase the server(s) (buddy and, if necessary, foe servers) taking part in administrating the logical file corresponding to array A are informed of the data structure characteristics and the declustering layout of the data. During the administration phase the buddy server of the application process tries to accomplish the request itself, otherwise it sends the request to the respective foe servers.

```
while write-requests(a) do
    D = Δ(a)
    for all d ∈ D
        if d ∈ BDL
            write O_d(a)
        else
            send write-request(a) to resp. foe server
        endif
    endfor
endwhile
```

The term *BDL* denotes the best-disk-list for each node (resp. server running on this node) as defined during the preparation phase (see Section 7.4) of the VIPIOS.

All write requests are performed in parallel on all servers. In the simplest case all data will be written locally. This can change, if a specified declustering scheme as decided during the administration phase chooses another disk.

---

[2] The VIPIOS_Put_Sec also provides other information in the argument list, which is omitted for simplicity.

Read array section. The reading of a file area corresponding to an array section is a little bit more complicated.

```
...
REAL  A(10, 1 : slabsize)
...
CALL  VIPIOS_Get_Sec(A, section_description, ... )
...
```

The procedure VIPIOS_Get_Sec executes a read request for the elements of the array A. This can be expressed by

---

```
SUBROUTINE VIPIOS_Get_Sec(A, section_description,  ... )
    contact buddy server
    ! let a be a file array area corresponding to the section of A
    which has to be read from the mass storage system
    read-request(a)
```

The requests are translated into a file area a according to the shape and the range of the requested regular section of A.

```
while read-requests(a) do
       D = Δ(a)
       for all d ∈ D do
            if d ∉ BDL
                 send read-request(a) to resp. foe server
            else ! d ∈ D !
                 B = read O_d(a)
                 Put(B)
            endif
       endfor
       for all d ∉ BDL
            B = receive O_d(a) from resp. foe server
            Put(b,p)
       endfor
endwhile
```

The procedure *Put* returns the read data to the requesting process, which can be application process or a foe server.

7.7 VIPIOS Servers - Structure and Communication Scheme

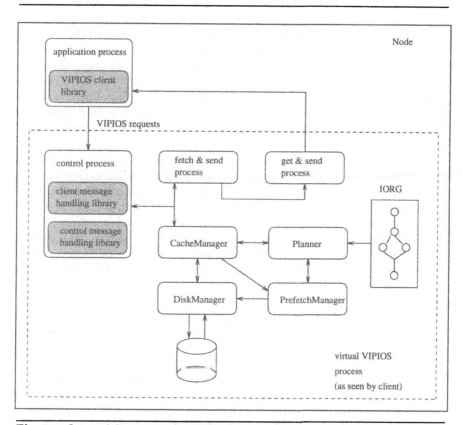

Fig. 7.8. Internal Structure of a VIPIOS Server

We shall now turn our attention to the design of VIPIOS servers (*VSs*) and on the communication structure within and among VSs with regard to the client application processes (*APs*). The VS components, called also functional units, and the relations among them are depicted in Fig. 7.8 which also shows a high-level view of the internal structure of a *VS* and the paths of communication between the units. The functional units are discussed below. This system is conceptually simple and is motivated by the internal structure of a parallel database system developed within the PARABASE project at the University of Vienna [196, 197].

In standard operation, each *AP* has to use a *client library*. This library provides access to the buddy server, located on the same node as the *AP* or

on a remote node. *AP* requests are passed to and results are obtained from
the buddy server.

We discuss the internal process structure of the *VS* first; the discussion
of the control flow and the data flow between the subprocesses and *AP*s in
case of write and read requests is postponed to subsequent parts.

From the technical point of view, each *VS* consists of seven processes,
control, fetch&send, get&send, CacheManager, DiskManager, Planner, and
PrefetchManager. Read requests are broadcasted to all operational *VS*s in
the system and subsequently are executed in parallel. The client library re-
siding in the *AP* has a counterpart in the *VS*, namely the *client message
handling library* used by the *VS control* process. The control process uses a
second message handling library, the control message handling library, for
all internal communications with other control processes on different nodes.
The *control* processes are responsible for coordinating tasks in the context
of parallel write and update operations as well as in the context of parallel
read operations. Additionally, directory handling and search is done by these
processes; it is necessary to break section requests down into requests for
data blocks. Retrieved data block numbers are passed to the corresponding
fetch&send processes, which are responsible for data block handling. Finally,
the corresponding *get&send* processes collect all subresults of the parallel
read operation from *fetch&send* processes, i.e., from the local as well as from
all other *fetch&send* processes on different *VS*s and pass the collected data
to client applications.

Remark: In case, *VS*s are placed on I/O nodes, it seems reasonable to move
get&send processes to *AP*s. Analysis of the corresponding performance issues
is one area of ongoing research.

Each *VS* has a buffer cache that is maintained by the CacheManager. The
cache is attached to a particular open file. In addition to deciding which
blocks are kept in the buffer, the CacheManager does all the work involved
in locating blocks in the buffer cache for fetch&send requests.

The *DiskManager* is responsible for actually reading data from and writing to disk. The DiskManager maintains a list of blocks that the CacheM-
anager has requested to be read or written. As new requests arrive from the
CacheManager, they are placed into the list according to the disk scheduling
algorithm.

It is helpful to prefetch data blocks into the cache before they are requested, so that the data is in the cache when requested. Without prefetching, the block is only read into the cache upon a cache miss, forcing the
process to experience a delay equal to the physical disk access time, plus any
queuing delays. Prefetching is attempted by a *VS* whenever the respective
application process is idle or dealing with a larger chunk of computation. The
prefetch module, called *PrefetchManager* running on the same *VS* repeatedly

considers prefetching, releasing control after each action. Each time, it calls a *Planner*, which makes its prediction based on the *IORG* information and other prediction heuristic. When asked for a prediction by the PrefetchManager, the Planner provides either a one- or more data block prediction, or chooses to make no prediction (sometimes, in this context, the best action is no action). If the predicted blocks are not already in the cache, the prefetching mechanism obtains free buffer locations and prefetches the blocks. The prefetch action is successful if and only if it issues a disk request.

7.7.1 Handling I/O Requests - Control Flow and Data Flow in the VIPIOS

In this part, we discuss control flow and data flow in the context of parallel *write*, *read*, and *prefetch* operations.

Write request. Each write request is passed to the control process of the buddy *VS* and triggers a lookup operation in the corresponding directory structure which yields the appropriate server numbers for the respective data blocks to be written. If one of the calculated server numbers refer to the buddy node, the data and the corresponding blocks numbers are passed to the fetch&send process and then finally written by this process.

If the calculated server number refers to a foe server, the control process passes the data to the corresponding control process, which takes the appropriate steps for local writing.

Read request. Control flow and data flow in the context of read operation executions represent probably the most interesting parts of the interprocess communication in the VIPIOS. Read requests can be handled in the similar way that query requests are handled by parallel database managers. The following description refers to the process structure discussed in Subsection 7.7 and to Fig. 7.9, which depicts the situation in case of a read operation execution; for the sake of simplicity, the Planner and PrefetchManager modules are omitted. This figure depicts the situation when *AP* and its buddy *VS* are located on the same node.

A particular read request on a node i is launched via an appropriate library call, received by a client message handling function and passed to the local control process on the node i. This responsible control process sends the query request to all other control processes (phase 1). All control processes perform an index search on their local part of the file in parallel and extract all relevant data block numbers from the directory (phase 2). These numbers are passed to the corresponding fetch&send processes. All fetch&send processes fetch the appropriate data blocks in parallel and send the retrieved data to the get&send process on the node i (phase 3). As soon as the get&send buffer area on node i is filled, the get&send process broadcasts some kind of "stop_transmission" signal to all fetch&send processes. Subsequently, it engages in the data delivery to the client process, i.e. the SPMD process

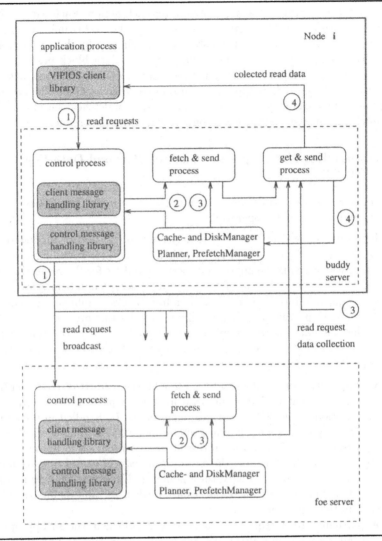

Fig. 7.9. Read Execution

(phase 4). As soon as the get&send buffer contents has been delivered, a "restart_transmission" signal is broadcasted by the get&send process and the receiving fetch&send processes restart their delivery operations. This protocol iterates in phase 3 and 4, until all the data qualified by the query request has been delivered.

Prefetch request. Control flow and data flow in the context of prefetch operation execution is almost the same as was discussed in the context of read request. However, during a prefetch request, the get&send process only collect the data for the I/O cache, not for the application process.

7.7.2 Implementation Notes

Basis of the VIPIOS implementation is the existing DiNG file system [196]. It is a prototype based on Distributed and Nested Grid files (i.e. DiNG file), which supports the efficient parallel execution of exact match, partial match and range queries directly by its inherent data structure. All necessary operations are provided at the system call level. Distributed and nested grid files are multikey index structures designed for mass-storage subsystems on distributed-memory architectures.

One of the existing opportunities is to implement the VIPIOS process and subprocess structure on top of the runtime interface called *Chant* [138] that has been developed at ICASE, NASA. Each functional unit is implemented as a separate, *lightweight, user level thread*. Lightweight threads offer significant advantages over heavyweight processing, including full control over thread scheduling, shared addressing spaces among threads within the same process, very fast context switching and the ability to execute on systems that do not provide multiprocess support [139].

Chant supports both a standardized interface for thread operations (as specified by the POSIX thread standard [151]) and communication among threads using either point-to-point primitives (such as those defined in the MPI standard [247]) or remote service requests (such as Active Messages [255]). Chant introduces the term *talking threads* to represent the notion of two threads in direct communication with each other, regardless of whether they exist in the same address space or not. A description of Chant, and its current status, can be found in [138].

BIBLIOGRAPHICAL NOTES

A lot of work is being carried out in many areas including devices, file systems, runtime techniques, algorithms, interface issues, and characterization of different I/O intensive applications. Earlier generations of parallel file systems like the Intel iPSC Concurrent File System (CFS) extended the Unix interface with file-pointer modes for parallel access [202, 212]. The Intel CFS was one of the first commercial parallel file systems that used file declustering over disks to improve I/O performance. Similar ideas were incorporated into the Thinking Machines' Connection Machine model CM-2's data vault [91]. File declustering was later implemented in many parallel file systems including the nCUBE Parallel I/O system [98], the Thinking Machines Scalable File System (SFS) [33], and the Intel Paragon Parallel File System [90]. SFS provides not only a Unix-like interface but has an additional collective I/O interface [33].

Among more recent parallel file systems, Vesta [85, 86] breaks away from the tradition of a one-dimensional file structure, and allows files to be viewed as two-dimensional entities where the user is capable of specifying both the logical partitioning of file data among the computational processors and the physical partitioning of file blocks across the disks. Vesta was developed as a research project, and led to IBM's Parallel I/O File System (PFIOS) [21], a commercial product.

The RAMA file system views the parallel file system as a block cache for the tertiary storage [190]; it distributes file blocks across disks randomly using a hash function.

In log-structured file systems [223], file system changes are buffered and periodically written to disk in a single, big disk-write operation, thereby speeding up both file writing and crash recovery.

Automatic caching and prefetching techniques for a BBN GP1000 multi-processor are investigated in [172].

The *Transparent Informed Prefetching* approach [210] focuses on extending the power of caching to reduce file read latencies by exploiting application level hints about future I/O accesses.

A dynamic prefetching scheme, called *ADOPT*, is described in [235]. This scheme exploits access patterns specified by the user or generated by the compiler.

Jovian [28, 29] is a parallel I/O runtime library supporting HPF applications. It restricts I/O to collective operations. In addition to application program processes, Jovian supports coalescing processes which filter application I/O requests and attempt to merge duplicate requests and join requests for adjacent data. The coalescing processes execute on the I/O nodes.

Stream* [191] is a parallel file abstraction for the data-parallel language C*.

Portable parallel file systems like PIOUS [195], and PPFS [149] have been developed recently. PIOUS is a parallel file system architecture for a network computing environment. The Portable Parallel File System (PPFS) [149] is a user-level library designed as an experimental testbed to study issues in parallel I/O. It is built above the standard Unix file system for portability to allow for testing on a variety of platforms. Interfaces allow the application to advertise intended access patterns and to control caching and prefetching.

The HFS file system designed as part of the Hurricane operating system, supports a wide variety of file structures and file system policies such as locking, prefetching, compression/decomposition and cache management [174].

Corbett et al. propose MPI-IO, which models I/O as message passing and allows programmers to express I/O with program datatypes rather than byte offsets within a file [247]. The goal of the MPI-IO proposal is to provide a standard for describing parallel I/O operations in MPI applications. MPI-IO uses MPI's concept of 'derived data types' to describe how data is laid out in a file.

The Parallel And Scalable Software for Input-Output (PASSION) project [80] at Syracuse is a broad effort to provide high performance parallel I/O at the language, compiler, runtime, and file system level. PASSION uses Two-phase I/O and its variants to obtain high performance out-of-core accesses [99]. In this approach, for read operations, the compute nodes cooperate to bring all the data into memory in a way that minimizes the total number of disk accesses by having the data layout in memory conform to the data layout on disk. The reverse strategy holds for writes. The two-phase strategy was extended to out-of-core computations in [246] to support reading and writing array sections.

Disk-directed I/O (DDIO) is an efficient implementation technique for collective I/O [170]. In DDIO, the computational processors make a collective I/O request, but the I/O processor dictates all the data transfer. The I/O processors are able to optimize performance by conforming I/O operations to both logical and physical file layouts. Under this approach, compute nodes tell the I/O nodes about a collective I/O request and, based on this semantic information, the I/O nodes direct the flow of data during the read or write operation. The I/O nodes direct the flow of data so they can form large continuous physical disk I/O requests rather than many smaller requests. The I/O nodes examine the request and determine which disk blocks will be accessed. The disk blocks can be accessed in sorted order, minimizing disk activity.

There are some disadvantages in implementing disk-directed I/O at the file system physical level, as intended by its creators. Chief among these is the introduction of operating system and file system dependences into implementations of disk-directed I/O, making ports difficult.

Nieuwejaar and Kotz propose a *nested-batched* interface for complex access patterns [201]. This interface has been implemented in Galley, a new parallel file system [200] using the DDIO technique. The Galley design has been guided by the results of several analyses of production file-system workloads on multiprocessors running scientific applications [220].

The PANDA runtime system is a server-directed I/O architecture that provides a high-level interface for array I/O [231]. It uses the DDIO optimization strategy.

Database research effort has focussed on the study of the use parallelism in the context of traditional DBMS operations, such as joins and range searching [101, 263]. Unfortunately, the vast of work on declustering relational data on parallel machines is not relevant for scientific data access patterns.

8. A New Generation Programming Environment for Parallel Architectures

8.1 Overview

In this chapter, we briefly discuss how we envision the further development of modern advanced programming environments for parallel computer systems. This discussion is based on research in the AURORA project [275] which aims to automate the selection of data distributions (in-core and out-of-core), exploit functional and data parallelism simultaneously, support performance tuning and high-level symbolic program debugging, and utilize expert system methods in the parallelization process.

Fig. 8.1 shows the main components of the AURORA compilation environment (ACE):

- the *restructuring system (VFCS+)*
  VFCS+ will be an extension of the VFCS and is the core of ACE. The following supporting tools will be included in VFCS+:
  - the *Vienna Fortran Performance Measurement System (VFPMS)*, advanced program measurement tool which allows the instrumentation of various code regions and collecting information for post-execution analysis
  - the P^3T, a parameter based-based performance estimator for prediction of communication and computation behavior of parallel programs
  - the *MIGRATOR*, a tool for automatic recognizing algorithmic concepts within a sequential code; it will also support the translation from a range of regular programs to HPF
  - a high-level symbolic debugger, called *Vienna Symbolic Debugger (VSD)* that will enable the Vienna Fortran/HPF+ programmer to observe the behavior of the program at the level at which the program has been developed
- the *performance analysis system (SCALA)*, a post execution performance and scalability analysis tool for parallel programs
- the *Knowledge-Based Support system (KBSS)* which will provide VFCS+ with decision support for the selection of restructuring strategies
- a graphical user interface provides a unique look and feel for the components of ACE.

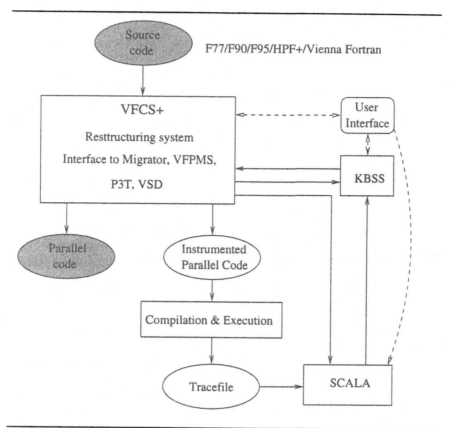

Fig. 8.1. AURORA Compilation Environment

Input languages to VFCS+ include Fortran 77 and Fortran 90/95 [189] as well as the variants of Vienna Fortran and HPF extensions called HPF+. [275].

8.2 Compilation and Runtime Technology

Fig. 8.2 gives an overview of those parts of VFCS+ which are directly associated with compile time transformations and the execution of parallel programs.

The source program is processed by the *front end*, normalized, and transformed into an intermediate representation which is kept in the *program database*. Program transformations are performed using compiler construction tools (PUMA[1]). The parallelization component transforms the data-parallel and task parallel Fortran program parts into an explicitly parallel pro-

---

[1] *PUMA* ia a component of the *Cocktail* compiler construction tool-set [134, 135].

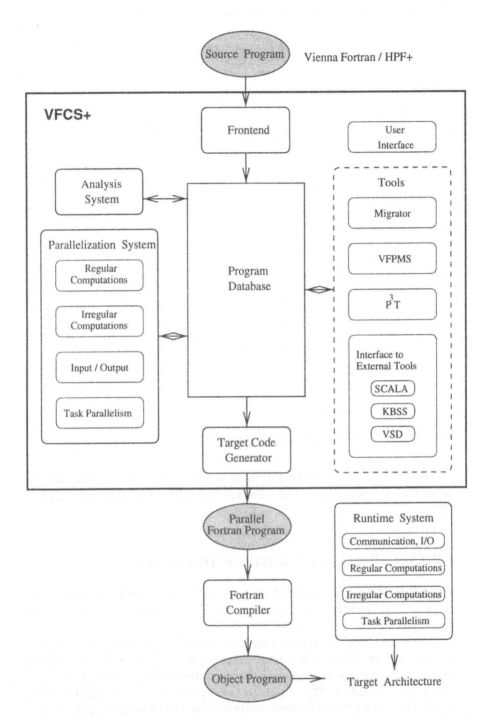

Fig. 8.2. VFCS+ Overview

gram. The transformation process handling the data-parallel program parts can conceptually be divided into three steps: (1) data distribution, (2) work distribution, and (3) communication and I/O generation. Based on extensive analyses powerful optimizations are performed to reduce communication and I/O costs. VFCS+ produces an explicitly parallel Fortran SPMD program, either based on the message-passing or on the shared memory paradigm.

VFCS+ can be used in an automatic or interactive mode. In the interactive mode, the user may receive feedback from the system, supply information to the system, and select transformation strategies.

The *runtime support system* will provide efficient mechanisms for I/O operations, handling runtime defined data and work distributions, irregular array accesses, and efficient methods for index conversion and static as well as dynamic schemes for handling the transfer of non-local data between the processors of the parallel target machine.

A range of analysis strategies will be developed in VFCS+, based on a generalization of the VFCS analysis component and of work done elsewhere [192, 193, 102, 103].

A unified data flow framework for program analysis will cover elimination of redundant and dead code, motion of loop-invariant code just before and just after the loop, and suppression of partially redundant and partially dead code in a uniform manner. Since the effectiveness of optimizations is increased by the accuracy of data flow information, the analysis will be performed interprocedurally.

This analysis framework will be applied to a broad range of problems, including dynamic data distribution, parallel I/O, irregular computations, communication optimizations, performance analysis, concept comprehension, and automatic data distribution. So, for instance, in the case of dynamically distributed arrays reaching distributions as well as redistribution placement has to be solved.

8.3 Debugging System

8.3.1 Motivation

Debugging is an integral part of the software development process. It enables the programmer to locate, analyze, and correct suspected faults. Debugging is, therefore, an interactive process which is based on the intuition of the programmer and her or his knowledge of the program to be investigated.

Due to this intuitive character, even debugging sequential programs involves many problems. Much greater difficulties occur in debugging parallel programs. A large gap exists between user needs and existing support for debugging parallel and distributed programs. The current approach is to perform debugging at the message-passing program level using a parallel debugger provided by the vendor of the underlying distributed-memory system. However, debugging at this abstraction level is difficult, because the user must exactly know how the program parallelized by a restructurer or compiler works. Especially, debugging of optimized message-passing code is a difficult problem. Without debugger support, it can be very tedious and time consuming, if not impractical. The user must be able to correlate the optimized code with the original source code. This is undesirable because it requires expert optimization knowledge and can be a challenging process even for an advanced compiler developer.

The design and implementation of a source-level debugging system that enables the Vienna Fortran/HPF+ programmer to observe the behavior of his program at the level at which the program has been developed presents unique challenges. To be practical, the debugger has to support interactive source-level debugging of large problems on large machines. It is then the task of the debugger to observe and control the state of many processors, to summarize and present distributed information in a concise and clear way in terms of the source program to the user.

Unfortunately, so far, only limited research investments have been put into the design and development of such high-level debugging tools. Prism [9], MDB [240], and PDT [81] are the only systems attempting to provide support for high-level debugging. The strength of Prism and PDT is their data visualization capability.

However, all of the debuggers mentioned above do not support debugging of optimized code; searching for errors in optimized code implies recompilation without optimizations and re-execution of the program. This may be impractical especially when processing large scale applications. For a realistic scenario, a program may need to be optimized before being able to execute or test it because of considerable execution time or problem size requirements. For example, optimization may significantly reduce the execution time to reach a point of program failure.

8.3.2 Design of a High-Level Symbolic Debugger

A high-level symbolic debugger, called VSD (Vienna Symbolic Debugger) will enable the Vienna Fortran/HPF+ programmer to observe the behavior of the program at the level at which the program has been developed.

We follow an approach referred to as *sequential view of parallel execution*; the real parallel code is executed, but a corresponding source code level interface is presented to the programmer. For example, the user is allowed to examine a distributed variable using the processor identification and the variable name occuring in the input program.

Portability of VSD is a design issue that is assigned high priority. The problems to be addressed can be summarized as follows:

- *Interface between VSD and compilation system:* By means of a portable interface VSD is able to provide the debugging support for various HPF compilation systems.
- *Interface between VSD and parallel machine:* The design relies on existing parallel debugger technology to implement the interface to parallel machines. If no vendor debugger is available for the target parallel machines, it is necessary to adapt a commonly available sequential debugger.

VSD offers all the features the user expects from a traditional source level debugger, such as the UNIX DBX debugger. Extra functionality is provided to aid in debugging data-parallel programs after an optimizing translation. This extra functionality includes

- Inspection of the values of an entire array or a section thereof.
- Built-in graphical data distribution visualization; this includes distribution of data on processors and external devices.
- Source-level debugging of optimized code.

Structure of the Debugging Support. The overall debugging support is split into three components: VSD, the Base Debugger (BD), and the BD/VSD communication interface. This is graphically sketched in Fig. 8.3. BD is a vendor-supplied debugger for the target parallel platform. It provides a collection of basic debugging functions. On the other hand, VSD provides more elaborate, higher level debugging functions built on top of the BD functionality. The interaction with the user is provided through a graphical interface (UI).

Interface Between the Compilation System and VSD. Any symbolic debugger attempts to present the user a picture of program execution that matches the source program. Specification of the correspondence between the source and object code is produced by the compiler and passed to the debugger at runtime.

VSD imposes the following requirements onto the underlying compiler via a runtime environment:

- a symbol table.
- relations between Vienna Fortran/HPF+ source code and parallel Fortran target code lines, as well as between variable names and types
- information about the transformations performed by the compiler.

Debugger Model. VSD uses the source browser to display source code. The user, for example, can set or remove breakpoints by clicking the mouse on the desired source line. VSD also accepts the debugging commands via its command line interface. The program output can be monitored in a separate display window while the program is executed on the target machine.

A Vienna Fortran / HPF+ program can be executed until the end of the program or a breakpoint has been reached. A breakpoint may be expressed

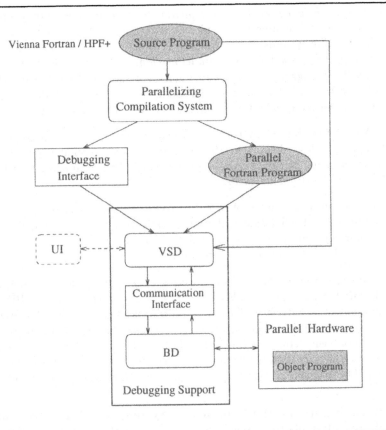

Fig. 8.3. Interaction of VSD with the Compilation and Runtime Environment

in terms of programming language control abstractions or data abstractions. Control breakpoints specify the break conditions in terms of the program's control flow (for example, "stop when line 530 is executed", or "stop when subroutine SUB1 is called"). Data breakpoints specify the break condition in terms of the program's memory state (for example, "stop when array A is modified or redistributed").

At a breakpoint, the user can inspect the value of a scalar variable or the values of a whole array or a section thereof. The values of specific variables or expressions can automatically be monitored at each breakpoint. Debugging commands enable the user to alter variable values.

Data visualization is a very important feature of VSD. VSD can visualize the distribution of arrays onto processors and disks. The user can navigate through the graphic representation of the regular or irregular distribution of an array section in order to reflect the quality of this distribution. By clicking on the graphical representation of an array element the user may obtain its associated value; it is also possible to alter the value of this element.

VSD also provides a convenient mechanism for viewing the values of entire arrays and array sections. An overview of data values is given that can be useful to detect illegal array elements which may cause program failures (such as a division by zero).

Advanced Features. Current implementations of all commercial symbolic debuggers including those for sequential programs rely heavily on non-optimized codes. There are various reasons for providing an HPF debugger that supports optimized codes:

- Even if the optimization system of the compiler is correct, there may exist program errors that are only revealed in an optimized version [264].
- For many HPF applications (e.g., irregular codes) debugging of an unoptimized code is not reasonable. For instance, it is not possible to execute and debug unoptimized out-of-core programs because of memory constraints.

The debugger has to convey the behavior of the optimized code in terms of the original code. However, when a program has been compiled with optimizations, mapping between breakpoints in the source and object code is complex, and values (distributions) of variables can be inaccessible at a breakpoint or inconsistent with the user's expectations. These problems have been pointed out by several designers of source-level debuggers of optimized sequential programs. So far, current research [1, 65, 83, 148, 264, 270, 276] offers partial solutions for a small class of optimizations, but not a unified approach that handles a wide range of optimizations.

There are two mutually exclusive classes of variables that are *endangered* by optimizations [1, 264]:

(i) *Noncurrent* variables. If at a breakpoint the value (distribution) of a variable V does not correspond to the value (distribution) that the user expects V to have in the source, V is said to be noncurrent at this breakpoint.
(ii) *Suspect* variables. There are situations when the debugger cannot tell whether the actual value (distribution) of V corresponds to the expected value of V. In this situation V is said to be a suspect variable.

In general, noncurrent and suspect variables are referred to as *endangered* variables. If a variable is not endangered, then it is *current*.

A debugger must detect and warn the user of noncurrent and suspect variables otherwise the user may draw incorrect conclusions about the program.

Fig. 8.4 shows a Vienna Fortran program section before and after code hoisting optimization. Using this example, we will demonstrate some of the problems a symbolic debugger for optimized code must deal with.

In Fig. 8.4, code hoisting has inserted the redistribution statement $D3$ into the ELSE-clause of the IF-statement, rendering the redistribution statement

```
D0:    REAL, DYNAMIC, DISTRIBUTED(BLOCK) :: A(N)
          ...
       IF (condition) THEN
D1:        DISTRIBUTE A :: (CYCLIC)
              ...
       ELSE

              ...
            breakpoint1
       END IF
       breakpoint2
D2:    DISTRIBUTE A :: (CYCLIC)
       ...
       breakpoint3
```

(a) Non-Optimized Version

```
D0:    REAL, DYNAMIC, DISTRIBUTED(BLOCK) :: A(N)
          ...
       IF (condition) THEN
D1:        DISTRIBUTE A :: (CYCLIC)
              ...
       ELSE
              ...
D3:        DISTRIBUTE A :: (CYCLIC)
            breakpoint1
       END IF
       breakpoint2
          ...
       breakpoint3
```

(b) Optimized Version

Fig. 8.4. Example of Code Hoisting

$D2$ redundant. At *breakpoint1*, array A is *noncurrent*, since the actual distribution of A is different from the expected distribution. If the user queries the distribution of A at this breakpoint, the debugger can display (visualize) the actual distribution of A and warn the user that this distribution has been obtained by the source-level statement $D2$, which has been executed earlier. The user may solve this problem by inserting a breakpoint before $D3$. At *breakpoint2*, A is *current* if execution has reached the breakpoint from the THEN-clause and *noncurrent* if execution has reached the breakpoint from the ELSE-clause. In the absence of knowledge regarding execution history, the debugger cannot determine whether A is current or noncurrent and consequently reports A as being suspect at *breakpoint2*. Note that the compiler can instrument the target code to collect runtime information, allowing the debugger to determine which control-flow path has been taken to *breakpoint2*. At *breakpoint3*, A is current because *D1* and *D3* yield the same distribution as given by *D2* in the non-optimized version.

VSD includes debugging techniques for detecting variables that are endangered by optimizations carried out during the parallelization of Vienna Fortran/HPF+ programs. In order to support this analysis, the compiler must perform bookkeeping to record the effects of optimizations in the internal program representation and pass the appropriate information to the debugger through a runtime environment. The debugging interface must be general enough to handle at least a large subset of common parallelizing transformations.

A similar approach was recently outlined in [1] for source-level debugging of scalar optimized code.

BIBLIOGRAPHICAL NOTES

Artificial Intelligence has been explicitly applied to the problem of generating codes for distributed-memory machines by Ko-Yang Wang at Purdue University [260]. He constructed a prototype system which includes descriptions of target machines, a set of program transformations, and rules for applying the transformations which are written in terms of the machine parameters.

The expert advisor EAVE [46] was developed to assist in the transformation of programs for input to the IBM 3090 VF.

The rule based transformation system ParTool [50] for semiautomatic SPMD program generation has been developed at the Delft University of Technology. ParTool enables the representation and reasoning about code transformation in the form of rewrite rules.

Andel, Hulman et al. [11] developed a pre-prototype implementation of an expert based support system that can either guide the user's choice of data distribution or generate it automatically within the context of an expert based parallelization environment.

Several tools have been developed and successfully applied to real applications that fall into the category of performance measurement/analysis tools: The MEasurements Description Evaluation and Analysis tool (MEDEA) [66] is a general-purpose environment which contains features for the analysis of performance data collected by measuring/monitoring tools.

The integration of the Pablo tool [2, 221] into the Fortran D compiler is an example on how performance debugging is done in the presence of a parallelizing compiler.

Fahringer [108] has designed and developed the prototype performance estimator $P^3 T$.

References

1. A. Adl-Tabatabai and T. Gross. Source-Level Debugging of Scalar Optimized Programs. *ACM SIGPLAN Notices. Proc. of the Conference on Programming Language Design and Implementation*, 31(5):178–189, May 1996.
2. V. S. Adve, J. Mellor-Crummey, M. Anderson, K. Kennedy, J. C. Wang, and D. A. Reed. An integrated compilation performance analysis environment for data parallel programs. In *Supercomputing '95, San Diego, CA, USA*, December 1995.
3. G. Agarwal and J. Saltz. Interprocedural compilation of irregular applications for distributed memory machines. In *Proceedings Supercomputing '95*, pages 785–796. IEEE Computer Society Press, December 1995.
4. G. Agarwal, J. Saltz, and R. Das. Interprocedural partial redundancy elimination and its application to distributed memory compilation. In *Proceedings of the SIGPLAN '95 Conference on Programming Language Design and Implementation*, pages 258–269. ACM Press, June 1995.
5. G. Agarwal, A. Sussman, and J. Saltz. Compiler and runtime support for structured and block structured applications. In *Proceedings of Supercomputing, Portland, OR*, pages 324–338, November 1993.
6. T. Agerwala, J.L. Martin, J.H. Mirza, D.C. Sadler, D.M. Dias, and M. Snir. SP2 system architecture. *IBM Systems Journal*, pages 152–184, 1995.
7. G. Agrawal, A. Acharya, and J. Saltz. An interprocedural framework for placement of asynchronous I/O operations. Technical Report, UMIACS and Department of Computer Science, University of Maryland, 1996.
8. A. V. Aho, R. Sethi, and J. D. Ullman. *Compilers: Principles, Techniques and Tools*. Addison-Wesley, 1985.
9. D. Allen, R. Bowker, K. Jourdenais, J. Simons, S. Sistare, and R. Title. The Prism Programming Environment. In *Proc. Supercomputer Debugging Workshop '91*, Albuquerque, NM, November 1991.
10. M. Alt, U. Assmann, and H. van Someren. CoSy compiler phase embedding with the CoSy compiler model. In *Proc. 5th International Conference on Compiler Construction*, pages 278–293, Edinburgh, U.K., April 1994. Springer-Verlag, LNCS 786.
11. S. Andel, B. M. Chapman, J. Hulman, and H. Zima. An expert advisor for parallel programming environments and its realization within the framework of the VFCS. Technical Report TR 95-4, Institute for Software Technology and Parallel Systems, University of Vienna, September 1995.
12. F. Andre, P. Brezany, O. Cheron, W. Dennisen, J. Pazat, K. Sanjari, , and E. van Konijnenburg. A new compiler technology for handling HPF data parallel constructs. In *Proceedings of the 3rd Workshop on Languages, and Run-Time Systems for Scalable Computers*, Troy, USA, May 1995. Kluwer.

13. F. Andre, J. Pazat, and H. Thomas. Pandore: A system to manage data distribution. In *ACM International Conference on Supercomputing, Amsterdam.* ACM Press, June 1991.

14. Raymond K. Asbury and David S. Scott. FORTRAN I/O on the iPSC/2: Is there read after write? In *Proceedings of the Fourth Conference on Hypercube Concurrent Computers and Applications*, pages 129–132. Golden Gate Enterprises, Los Altos, CA, 1989.

15. G. Bachler and R. Greimel. Parallel CFD in the industrial environment. Unicom Seminars, London, 1994.

16. H. E. Bal, M. F. Kaashoek, and A. S. Tanenbaum. Orca: A language for parallel programming of distributed systems. *IEEE Transactions on Software Engineering*, 18(3):190–205, March 1992.

17. P. Banerjee et al. The Paradigm compiler for distributed-memory multicomputers. *IEEE Transactions on Computers*, 28(10):37–47, October 1995.

18. H. Bao et al. Earthquake ground motion modeling on parallel computers. In *Proc. Supercomputing' 96, Pittsburgh.* IEEE Comp. Soc. Press, November 1996.

19. S. T. Barnard and H. Simon. A fast multilevel implementation of recursive spectral bisection for partitioning unstructured problems. Technical Report RNR-92-033, NAS Systems Division, NASA Ames Research Center, November 1992.

20. R. Barrett, M. Berry, T. Chan, J. Demmel, J. Donato, J. Dongarra, V. Eijkhout, R. Pozo, C. Romine, and H. van der Vorst. *Templates for the solution of linear systems: Building blocks for iterative methods.* SIAM, 1994.

21. Fern E. Bassow. Installing, managing, and using the IBM AIX Parallel I/O File System. IBM Document Number SH34-6065-00, February 1995. IBM Kingston, NY.

22. S. Benkner. Vienna Fortran 90 and its compilation. Technical Report TR-94-8 (also PhD Thesis), Institute for Software Technology and Parallel Systems, University of Vienna, November 1994.

23. S. Benkner, P. Brezany, and H. P. Zima. Functional specification of the PREPARE Parallelization Engine. Technical Report Vienna-9005-PESpec (PREPARE Project Document), Institute for Software Technology and Parallel Systems, University of Vienna, June 1993.

24. S. Benkner, P. Brezany, and H. P. Zima. Program normalization. Technical Report Vienna-9006-norm (PREPARE Project Document), Institute for Software Technology and Parallel Systems, University of Vienna, June 1993.

25. S. Benkner, P. Brezany, and H. P. Zima. Processing array statements and procedure interfaces in the Prepare High Performance Fortran compiler. In *Proc. 5th International Conference on Compiler Construction*, pages 324–338, Edinburgh, U.K., April 1994. Springer-Verlag, LNCS 786.

26. S. Benkner, B. Chapman, and H.P. Zima. Vienna Fortran 90. *Proceeding of the Scalable High Perfomance Computing Conference*, pages 51–49, April 1992.

27. S. Benkner et al. *Vienna Fortran Compilation System. User's Guide*, version 2.0 edition, October 1995.

28. R. Bennent, Kelvin Bryant, Alan Sussman, Raja Das, and Joel Saltz. Jovian: A framework for optimizing parallel I/O. *IEEE Computer Society Press*, October 1994.

29. R. Bennent et al. A framework for optimizing parallel I/O. *Submitted to ICS'95*, 1995.

30. A. Bequelin et al. A user's guide to PVM: Parallel virtual machine. Technical Report TR TM-11826, Oak Ridge National Laboratory, 1991.

31. M. J. Berger and S. H. Bokhari. A partitioning strategy for nonuniform problems on multiprocessors. *IEEE Trans. on Computers*, C-36(5):570–580, May 1987.

32. R. Berrendorf, M. Gerndt, W. Nagel, and J. Pruemmer. Fortran. Technical Report KFA-ZAM-IB-9322, Forschungszentrum Juellich GmbH, Germany, November 1993.

33. Michael L. Best, Adam Greenberg, Craig Stanfill, and Lewis W. Tucker. CMMD I/O: A parallel Unix I/O. In *Proceedings of the Seventh International Parallel Processing Symposium*, pages 489–495, 1993.

34. F. Bodin, L. Kervella, and T. Priol. Fortran S: A Fortran interface for shared virtual memory architectures. In *Proceedings of Supercomputing 1993, Portland, USA*. IEEE, 1993.

35. S. H. Bokhari. Partitioning problems in parallel, pipelined, and distributed computing. *IEEE Transactions on Computers*, 37(1):48–57, January 1988.

36. L. Bomans, D. Roose, and R. Hempel. The Argonne/GMD macros in Fortran for portable parallel programming and their implementation on the Intel iPSC/2. Technical report, Elsevier Science, 1990.

37. R. Bordawekar, A. Choudhary, and J. M. Rosario. An experimental performance evaluation of Touchstone Delta concurrent file system. In *Proc. Int. Conf. on Supercomputing*, July 1993.

38. R. Bordawekar, J. M. Rosario, and A. Choudhary. Design and evaluation of primitives for parallel I/O. In *Proc. Supercomputing '93*, November 1993.

39. R. R. Bordawekar and A. N. Choudhary. Language and compiler support for parallel I/O. In *IFIP Working Conference on Programming Environments for Massively Parallel Distributed Systems*. Swiss, April 1994.

40. Rajesh Bordawekar. *Techniques for Compiling I/O Intensive Parallel Programs*. PhD thesis, Department of Computer and Information Science, Syracuse University, Syracuse, NY, April 1994.

41. Rajesh Bordawekar and Alok Choudhary. Compiler and runtime support for parallel I/O. In *Proceedings of IFIP Working Conference (WG10.3) on Programming Environments for Massively Parallel Distributed Systems*, Monte Verita, Ascona, Switzerland, April 1994. Birkhaeuser Verlag AG, Basel, Switzerland.

42. Rajesh Bordawekar and Alok Choudhary. Issues in compiling I/O intensive problems. In Ravi Jain, John Werth, and James C. Browne, editors, *Input/Output in Parallel and Distributed Computer Systems*, chapter 3, pages 69–96. Kluwer Academic Publishers, 1996.

43. Rajesh Bordawekar, Alok Choudhary, Ken Kennedy, Charles Koelbel, and Michael Paleczny. A model and compilation strategy for out-of-core data parallel programs. In *Proceedings of the Fifth ACM SIGPLAN Symposium on Principles and Practice of Parallel Programming*, pages 1–10, July 1995.

44. Rajesh Bordawekar, Alok Choudhary, and J . Ramanujam. Automatic optimization of communication in compiling out-of-core stencil codes. Technical Report CACR–114, Center for Advanced Computing Research, California Institute of Technology, November 1995.

45. R.R. Bordawekar and A.N. Choudhary. Communication strategies for out-of-core programs on distributed memory machines. In *Proceedings the ICS'95*, pages 395–403, July 1995.

46. P. Bose. Interactive program improvement via EAVE: an expert adviser for vectorization. In *International Conference on Supercomputing*, pages 119–130, July 1988.

47. Z. Bozkus, A. Choudhary, G. Fox, T. Haupt, and S. Ranka. Fortran 90D/HPF compiler for distributed-memory MIMD computers: Design, implementation,

and performance results. In *Proceedings of the 1993 ACM International Conference on Supercomputing.* ACM Press, July 1993.

48. David K. Bradley and Daniel A. Reed. Performance of the Intel iPSC/2 input/output system. In *Proceedings of the Fourth Conference on Hypercube Concurrent Computers and Applications*, pages 141–144. Golden Gate Enterprises, Los Altos, CA, 1989.

49. T. Brandes. Efficient data parallel programming without explicit message passing distribute memory multiprocessors. Technical Report AHR-92-4, GMD, 1992.

50. L. C. Breebaart, E. M. Paalvast, and H. J. Sips. A rule based transformation system for parallel languages. In *Third Workshop on Compilers for Parallel Computers*, pages 13–21. Vienna, Austria, ACPS/TR 93-8, July 1992.

51. P. Brezany. Denotational definition of Vienna Fortran I/O operations. Research Report of the VIPIOS Project, Institute for Software Technology and Parallel Systems, University of Vienna, Austria, March 1997.

52. P. Brezany, B. Chapman, R. Ponnusamy, V. Sipkova, and H. Zima. Study of application algorithms with irregular distributions. Technical Report D1Z-3 of the CEI-PACT Project, University of Vienna, April 1994.

53. P. Brezany, O. Cheron, K. Sanjari, and E. van Konijnenburg. Procesing irregular codes containing arrays with multi-dimensional distributions by the PREPARE HPF compiler. In *Proceedings of the Conference "High Performance Computing and Networking 1995 Europe"*, pages 526–531, Milano, Italy, Springer-Verlag, LNCS 919, May 1995.

54. P. Brezany, M. Dang, and E. Schikuta. Handling standard Fortran 90 I/O operations by the parallelizing compiler and parallel I/O runtime system. Research Report of the VIPIOS Project, Institute for Software Technology and Parallel Systems, University of Vienna, Austria, March 1997.

55. P. Brezany, M. Gerndt, P. Mehrotra, and H. Zima. Concurrent file operations in a High Performance FORTRAN. In *Proceedings of Supercomputing '92*, pages 230–237, 1992.

56. P. Brezany, M. Gerndt, and V. Sipkova. SVM Support in the Vienna Fortran Compilation System. Technical Report KFA ZAM-IB-9401, Forschungszentrum Juelich, Germany, March 1994.

57. P. Brezany, M. Gerndt, V. Sipkova, and H. Zima. SUPERB support for irregular scientific computations. In *Proc. of the Scalable High Performance Computing Conference*, pages 314–321, Williamsburg, USA, April 1992.

58. P. Brezany and J. Knoop. Interprocedural optimizations of irregular applications for distributed-memory systems. in preparation, 1997.

59. P. Brezany, T. A. Mueck, and E. Schikuta. Language, compiler and advanced data structures support for parallel I/O operations. Internal Report of the Institute for Software Technology and Parallel Systems, December 1994.

60. P. Brezany, T. A. Mueck, and E. Schikuta. Automatic parallelization of Vienna Fortran / HPF programs operating on huge data structures. Internal Report of the Institute for Software Technology and Parallel Systems, July 1996.

61. P. Brezany and V. Sipkova. Implementation of indirect distributions. Technical Report of the PPPE project, Institute for Software Technology and Parallel Systems, University of Vienna, September 1995.

62. P. Brezany and V. Sipkova. Coupling parallel data and work partitioners to the Vienna Fortran Compilation System. In *Proceedings of the Conference EUROSIM – HPCN Challenges 1996*. North Holland, Elsevier, June 1996.

63. P. Brezany and V. Sipkova. Handling HPF/Vienna Fortran irregular codes by the Vienna Fortran Compilation Systems. In Preparation, April 1997.

64. P. Brezany, V. Sipkova, B. Chapman, and R. Greimel. Automatic parallelization of the AVL FIRE benchmark for a distributed-memory system. In *Proceedings of the Conference PARA '95*, LNCS Springer-Verlag, August 1995.

65. G. Brooks, G.J. Hansen, and S. Simons. A New Approach to Debugging Optimized Code. *ACM SIGPLAN Notices. Proc. of the Conference on Programming Language Design and Implementation*, 27(7):1–11, July 1992.

66. M. Calzarossa, L. Massari, A. Merlo, M. Pantano, and D. Tessera. MEDEA: A tool for workload chracterization of parallel systems. *IEEE Parallel and Distributed Technology*, pages 72–80, 1995.

67. A. Carle, M. Hall, J. Mellor-Crummey, and R. Rodriguez. FIAT: A framework for interprocedural analysis and transformation. Technical Report CRPC-TR95522-S, Rice University, March 1995.

68. D. B. Carpenter. Array communication library: User guide and reference manual, May 1991.

69. N. Carriero and D. Gelenter. *How to Write Parallel Programs*. MIT Press, 1990.

70. L. Carter, J. Ferrante, and S. Flynn Hummel. Hierarchical tiling for improved superscalar performance. *IPPS Conference*, April 1995.

71. B. M. Chapman, P. Mehrotra, J. van Rosendale, and H. Zima. A software architecture for multidisciplinary applications: Integrating task and data parallelism. In *Proceedings of the CONPAR'94 Conference, Linz, Austria*, pages 664–676, September 1994.

72. B. M. Chapman, P. Mehrotra, and H. P. Zima. Programming in Vienna Fortran. *Scientific Programming*, 1:31–50, Fall 1992.

73. B. M. Chapman, P. Mehrotra, and H. P. Zima. Extending HPF for advanced data parallel applications. Technical Report TR 94-7, Institute for Software Technology and Parallel Systems, University of Vienna, 1994.

74. B. M. Chapman et al. *Vienna Fortran Compilation System. User's Guide*, January 1993.

75. M. Chen and J. Li. Index domain alignment: Minimizing cost of cross-referencing between distributed arrays. In *Frontiers90 of the 3rd Symposium on the Frontiers of Massively Parallel Computation*. College Park, MD, October 1990.

76. Peter M. Chen, Edward K. Lee, Garth A. Gibson, Randy H. Katz, and David A. Patterson. RAID: high-performance, reliable secondary storage. *ACM Computing Surveys*, 26(2):145–185, June 1994.

77. A. Choudhary, C. Koelbel, and K. Kennedy. Preliminary proposal to provide support for OOC arrays in HPF, September 1995.

78. Alok Choudhary, Rajesh Bordawekar, Michael Harry, Rakesh Krishnaiyer, Ravi Ponnusamy, Tarvinder Singh, and Rajeev Thakur. PASSION: parallel and scalable software for input-output. Technical Report SCCS-636, ECE Dept., NPAC and CASE Center, Syracuse University, September 1994.

79. Alok Choudhary, Rajesh Bordawekar, Sachin More, K. Sivaram, and Rajeev Thakur. PASSION runtime library for the Intel Paragon. In *Proceedings of the Intel Supercomputer User's Group Conference*, June 1995.

80. A. Choudhary et al. PASSION: Parallel and scalable software for input-output. Technical Report CRPC-TR94483, Rice University, Houston, 1994.

81. C. Clemenson, J. Fritscher, and R. Ruehl. Visualization, Execution Control and Replay of Massively Parallel Programs within Annai's Debugging Tool. CSCS TR-94-09, CSCS ETH, Manno, Switzerland, November 1994.

82. Fabien Coelho. Compilation of I/O communications for HPF. In *Proceedings of the Fifth Symposium on the Frontiers of Massively Parallel Computation*, pages 102–109, 1995.

83. R. Cohn. Source Level Debugging of Automatically Parallelized Code. In *In Proc. ACM/ONR Workshop on Parallel and Distributed Debugging*, pages 128–139, Santa Cruz, CA, May 1991.

84. PREPARE Consortium. Design of advanced PARTI schedule reuse optimizer. Technical Report GMD-5006-3 (PREPARE Project Document), GMD FIRST Berlin, March 1996.

85. P. F. Corbett and D. G. Feitelson. Design and implementation of the Vesta parallel file system. In *Proc. Scalable High Performance Computing Conference*, May 1994.

86. Peter F. Corbett, Dror G. Feitelson, Jean-Pierre Prost, George S. Almasi, Sandra Johnson Baylor, Anthony S. Bolmarcich, Yarsun Hsu, Julian Satran, Marc Snir, Robert Colao, Brian Herr, Joseph Kavaky, Thomas R. Morgan, and Anthony Zlotek. Parallel file systems for the IBM SP computers. *IBM Systems Journal*, pages 222–248, 1995.

87. P. Corbett et al. MPI-IO: A parallel file I/O interface for MPI. version 0.3. Technical Report NAS-95-002, NAS, January 1995.

88. Thomas H. Cormen and Alex Colvin. ViC*: A preprocessor for virtual-memory C*. Technical Report PCS-TR94-243, Dept. of Computer Science, Dartmouth College, November 1994.

89. Intel Corporation. iPSC/2 and iPSC/860 manuals, 1990.

90. Intel Corporation. Paragon XP/S product overview. Technical report, Intel Corporation, 1991.

91. Thinking Machines Corporation. Connection Machine model CM-2 technical summary. Technical Report HA87-4, Thinking Machines, April 1987.

92. Thinking Machines Corporation. *CM Fortran Reference Manual. Version 5.2*, April 1989.

93. D. Culler, R. Karp, D. Patterson, A. Sahay, K. E. Schauser, E. Santos, R. Subramonian, and T. von Eicken. LogP: Towards a realistic model of parallel computation. In *Proceedings of 4th ACM Symposium on Principles and Practice of Parallel Programming*, pages 1–12, 1993.

94. J. Darlington, Y. Guo, H. W. To, and J. Yang. Parallel skeletons for structured composition. In *PPOPP'95, Santa Clara, CA, USA*, pages 19–28, 1995.

95. R. Das and J. Saltz. *A manual for PARTI runtime primitives - Revision 2*, December 1990.

96. R. Das, J. Saltz, and R. von Hanxleden. Slicing analysis and indirect accesses to distributed arrays. Technical Report CS-TR-3076, UMIACS-TR-93-42, University of Maryland, College Park, MD, 1993.

97. R. Das, A. Sussman, P. Havlak, and J. Saltz. Compiler analysis and optimization of indirect array accesses. In *5th International Workshop on Compilers for Parallel Computers, Malaga, Spain*, pages 98–110, June 1995.

98. E. DeBenedictis and J. M. del Rosario. nCUBE parallel I/O software. In *11th Annual IEEE Int. Phoenix Conf. on Computers and Communications*, pages 117–124, April 1992.

99. Juan Miguel del Rosario, Rajesh Bordawekar, and Alok Choudhary. Improved parallel I/O via a two-phase run-time access strategy. In *IPPS '93 Workshop on Input/Output in Parallel Computer Systems*, pages 56–70, 1993. Also published in Computer Architecture News 21(5), December 1993, pages 31–38.

100. Juan Miguel del Rosario and Alok Choudhary. High performance I/O for parallel computers: Problems and prospects. *IEEE Computer*, 27(3):59–68, March 1994.

101. D. DeWitt and J. Gray. Parallel database systems: The future of high performance database systems. *Communications of the ACM*, 6(35):85–98, June 1992.

102. D. M. Dhamdhere. A fast algorithm for code movement optimisation. *SIG-PLAN Notices*, 23:75–93, October 1988.
103. D. M. Dhamdhere and H. Patil. An elimination algorithm for bidirectional data flow problems using edge placement. *ACM Transactions on Programming Languages and Systems*, 15:312–336, April 1993.
104. D. M. Dhamdhere, B. K. Rosen, and F. K. Zadeck. How to analyze large programs efficiently and informatively. In *Proceedings of the SIGPLAN '92 Conference on Programming Language Design and Implementation*, pages 212–223. ACM Press, June 1992.
105. J. Docter, K. Falski, and J. Knecht. Paragon XP/S at KFA Juelich. Technical Report KFA-ZAM-TKI-0230, Forschungszentrum Juelich, Germany, February 1996.
106. R. J. Duffin. On Fourier's analysis of linear inequality systems. *Mathematical Programming Study*, 1:71–95, 1974.
107. J. Erxleben. Development of parallel components for a parallel file system of a massively parallel computer (in german). Technical Report Master Thesis, Technical University Chemnitz - Zwickau, Germany, 1993.
108. T. Fahringer and H. Zima. A static parameter based performance prediction tool for parallel programs. Technical Report ACPC/TR 93-1, Vienna University, January 1993.
109. N. Floros and J. Reev. Domain decomposition tool (DDT). Technical Report Esprit CAMAS 6756, University of Southampton, March 1994.
110. High Performance Fortran Forum. High performance Fortran language specification. *Scientific Programming*, 2(1,2):1–100, Januar 1993.
111. High Performance Fortran Forum. Hpf-2 scope of activities and motivating applications. version 0.8, November 1994.
112. Message Passing Interface Forum. MPI: A message-passing interface standard, April 1994.
113. I. Foster. *Designing and building parallel programs*. Addison-Wesley, 1995.
114. I. Foster and K. M. Chandi. Fortran M: A language for modular parallel programming. *Journal on Parallel and Distributed Computing*, 25(1), January 1995.
115. G. Fox et al. *Solving Problems on Concurrent Processors*. Prentice Hall, 1988.
116. G. Fox et al. Fortran D language specification. Technical Report COMP TR90-14, Rice University, December 1990.
117. N. Galbreath, W. Gropp, and D. Levine. Applications-driven parallel I/O. In *Proceedings of Supercomputing '93*, pages 462–471, 1993.
118. H. M. Gerndt. *Automatic Parallelization for Distributed-Memory Multiprocessing Systems*. PhD thesis, University of Bonn, 1989.
119. H. M. Gerndt. Work distribution in parallel programs for distributed memory multiprocessors. In *Proceedings of the International Conference on Supercomputing*, pages 96–103. CACM, July 1991.
120. H. M. Gerndt and H. P. Zima. Optimizing communication in SUPERB. In *Procedings of CONPAR90*, 1990.
121. Garth A. Gibson. *Redundant Disk Arrays: Reliable, Parallel Secondary Storage*. ACM Distinguished Dissertations. MIT Press, 1992.
122. W. K. Giloi. Parallel programming models and their interdependence with parallel architectures. In *Proceedings of the Conference: Programming Models for Massively Parallel Computers*, pages 2–11, Berlin, September 1993. IEEE Computer Society Press.
123. W. K. Giloi. *Rechnerarchitektur*. Springer-Lehrbuch. Springer-Verlag, 1993.

124. W. K. Giloi, S. Jaehnichen, and B. D. Shriver (eds.). *Programming Models for Massively Parallel Computers*. Proceedings of the Conference. IEEE Computer Society Press, 1993.

125. W. K. Giloi, S. Jaehnichen, and B. D. Shriver (eds.). *Programming Models for Massively Parallel Computers*. Proceedings of the Conference. IEEE Computer Society Press, 1995.

126. R. Goering. The IBM 9076 scalable powerparallel SP2 system: System management. IBM Kingston, April 27, 1994. Draft.

127. E. H. Gornish. Compile time analysis for data prefetching. Technical Report CSRD R-949, University of Illinois, Urbana, December 1989.

128. E. H. Gornish. *Adaptive and Integrated Data Cache Prefetching for Shared-Memory Multiprocessors*. PhD thesis, CSRD, University of Illinois, Urbana, 1995.

129. E. H. Gornish, E. Granston, and A. Veidenbaum. Compiler-directed data prefetching in multiprocessors with memory hierarchies. In *Proceedings of the International Conference on Supercomputing*, pages 354–368. ACM, 1990.

130. A. Goscinski. *Distributed Operating Systems. The Logical Design*. Addison-Wesley, 1992.

131. J. Griffien and R. Appleton. Reducing file system latency using a predictive approach. In *Proceedings of the 1994 Summer USENIX Conference*. USENIX, 1994.

132. W. Gropp, E. Lusk, and A. Skjellum. *Using MPI: Portable Parallel Programming with the Message Passing Interface*. MIT Press, 1994.

133. W. Gropp and B. Smith. Chameleon parallel programming tools user's manual. Technical Report TR ANL-93/23, Argonne National Laboratory, March 1993.

134. J. Grosch. Puma - a generator for the transformation of attributed trees, compiler generation. Report No.26, GMD Forschungsstelle an der Universität Karlsruhe, Germany, November 1988.

135. J. Grosch. Ast - a generator for abstract syntax tree. Report No. 15, GMD Forschungsstelle an der Universität Karlsruhe, Germany, September 1991.

136. M. Grossman. Modeling reality. *IEEE Spectrum*, 29(9):56–60, September 1992.

137. M. Gupta, E. Schonberg, and H. Srinivasan. A unified framework for optimizing communication in data-parallel programs. In *Proceedings of the 7th Workshop on Languages and Compilers for Parallel Computing*, pages 334–345. Springer-Verlag, August 1994.

138. M. Haines, D. Cronk, and P. Mehrotra. On the Design of Chant: A Talking Threads Package. In *Proceedings of Supercomputing 94, Washington, D.C.*, pages 350–359, November 1994.

139. M. Haines, B. Hess, P. Mehrotra, J. Van Rosendale, and H. P. Zima. Runtime support for data parallel tasks. Technical Report TR 94-2, Institute for Software Technology and Parallel Systems, University of Vienna, April 1994.

140. M. W. Hall, S. Hiranandani, K. Kennedy, and C. Tseng. Interprocedural compilation of Fortran D for MIMD distributed-memory machines. In *Supercomputing'92, Minneapolis*, pages 522–534, November 1992.

141. P. J. Hatcher and M. J. Quinn. *Data-Parallel Programming on MIMD architectures*. MIT Press Scientific and Engineering Computation Series, 1991.

142. M. S. Hecht. *Flow Analysis of Computer Programs*. North-Holland, 1977.

143. High Performance Fortran Forum. High Performance Fortran. Version 2.0.δ, October 1996.

144. W. Daniel Hillis and Lewis W. Tucker. The CM-5 connection machine: A scalable supercomputer. *Communications of the ACM*, 36(11):31–40, November 1993.

145. S. Hiranandani, K. Kennedy, and C. Tseng. Compiler optimizations for Fortran D on MIMD distributed-memory machines. In *Proceedings of the Supercomputing Conference 1991*, pages 86–100, November 1991.

146. S. Hiranandani, K. Kennedy, and C. W. Tseng. Compiling Fortran D for MIMD distributed-memory machines. *Communications of the ACM*, 35(8):66–80, August 1992.

147. C. A. R. Hoare. Communicating sequential processes. *Communications of the ACM*, 21(8):666–677, August 1978.

148. U. Hoelzle, C. Chambers, and D. Ungar. Debugging Optimized Code with Dynamic Deoptimization. *ACM SIGPLAN Notices. Proc. of the Conference on Programming Language Design and Implementation*, 27(7):32–43, July 1992.

149. J. Huber, C. Elford, D. Read, A. Chien, and D. Blumenthal. PPFS: a high performance portable parallel file system. In *Proceedings of the 9th ACM International Conference on Supercomputing*, pages 385–394, July 1995.

150. Yuan-Shin Hwang, Bongki Moon, Shamik D. Sharma, Ravi Ponnusamy, Raja Das, and Joel H. Saltz. Runtime and language support for compiling adaptive irregular programs. *Software Practice and Experiments*, 25(6):597–621, June 1995.

151. IEEE. Threads extension for portable operating systems (version 7), February 1992.

152. Applied Parallel Research Inc. APR's FORGE 90 parallelization tools for High Performance Fortran. Internal Report, June 1993.

153. J. M. del Rosario and A. N. Choudhary. High performance I/O for parallel computers: Problems and prospects. *IEEE Computer*, pages 59–68, March 1994.

154. H. F. Jordan. Scalability of data transport. In *Proc. of the Scalable High Performance Computing Conference*, pages 1–8, Williamsburg, USA, April 1992.

155. M. Kandemir, R. Bordawekar, and A. Choudhary. Data access reorganizations in compiling out-of-core data parallel programs on distributed memory machines, November 1996.

156. A. Karp. Programming for parallelism. *IEEE Computer*, 20(5):43–57, January 1987.

157. Ken Kennedy and Nenad Nedeljkovic. Combining dependence and data-flow analyses to optimize communication. Technical Report CRPC-TR94484-S, Rice University, September 1994.

158. F. Kim. UniTree: A closer look at solving the data storage problem, September 1996.

159. Michelle Y. Kim. Synchronized disk interleaving. *IEEE Transactions on Computers*, C-35(11):978–988, November 1986.

160. J. Knoop. A framework for interprocedural data flow analysis with applications in assignment sinking and hoisting, November 1996.

161. J. Knoop, O. Ruething, and B. Steffen. Lazy code motion. In *Proceedings of the SIGPLAN '92 Conference on Programming Language Design and Implementation*, pages 224–234. ACM Press, June 1992.

162. J. Knoop, O. Ruething, and B. Steffen. Partial dead code elimination. In *Proceedings of the SIGPLAN '94 Conference on Programming Language Design and Implementation*, pages 147–158. ACM Press, June 1994.

163. J. Knoop, O. Ruething, and B. Steffen. Towards a tool kit for the automatic generation of interprocedural data flow analyses, November 1994.

164. J. Knoop, O. Ruething, and B. Steffen. The power of assignment motion. In *Proceedings of the SIGPLAN '95 Conference on Programming Language Design and Implementation*, pages 233–245. ACM Press, June 1995.

165. C. Koelbel. *Compiling Programs for Nonshared Memory Machines*. PhD thesis, Purdue University, West Lafayette, November 1990.

166. C. Koelbel. Compile time generation of regular communications patterns. In *Proc. Supercomputing 91*, pages 101–110, Albuquerque, 1991.

167. C. Koelbel, P. Mehrotra, and J. Van Rosendale. Supporting shared data structures on distributed memory architectures. In *2nd ACM SIGPLAN Symposium on Principles Practice of Parallel Programming*, pages 177–186. CACM, March 1990.

168. C. H. Koelbel et al. *The High Performance Fortran Handbook*. Scientific and Engineering Computation Series. MIT Press, 1994.

169. S. R. Kohn and S. B. Baden. A robust parallel programming model for dynamic non-uniform scientific computations. In *Proceedings of SHPCC, Knoxville*. IEEE Press, May 1994.

170. D. Kotz. Disk-directed I/O for MIMD multiprocessors. Technical Report PCS-TR94-226, Department of Computer Science, Darmouth College, July 1994.

171. D. Kotz. Applications of parallel I/O. Technical Report PCS-TR96-297, Department of Computer Science, Darmouth College, October 1996.

172. David Kotz. *Prefetching and Caching Techniques in File Systems for MIMD Multiprocessors*. PhD thesis, Duke University, April 1991. Available as technical report CS-1991-016.

173. David Kotz. Introduction to multiprocessor I/O architecture. In Ravi Jain, John Werth, and J. C. Browne, editors, *Input/Output in Parallel and Distributed Computer Systems*, pages 97–123. Kluwer Academic Publishers, 1996.

174. Orran Krieger. *HFS: A flexible file system for shared-memory multiprocessors*. PhD thesis, University of Toronto, October 1994.

175. H. W. Kuhn. Solvability and consistency for linear equations and inequalities. *American Mathematical Monthly*, 63:217–232, 1956.

176. V. Kumar, A. Grama, A. Gupta, and G. Karypis. *Introduction to Parallel Computing*. The Benjamin/Cummings Publishing Company, 1994.

177. T. T. Kwan and D. A. Reed. Performance of the CM-5 scalable file system. In *Proceedings of the 8th ACM International Conference on Supercomputing*, pages 156–165, July 1994.

178. F. Langhammer and F. Wray. Supercomputing and transputers. In *International Conference on SUPERCOMPUTING*, pages 114–128. CACM, July 1992.

179. F. F. Lee. Partitioning of regular computation on multiprocessor systems. *IEEE Transactions on Parallel and Distributed Systems*, 1(9):312–317, September 1990.

180. C. E. Leiserson et al. The network architecture of the Connection Machine CM-5. In *ACM Symposium on Parallel Algorithms and Architectures*. CACM, July 1992.

181. Li and Fuchs. Catch - compiler-assisted techniques for checkpointing. In *Proc. of the IEEE Symposium on Fault-Tolerant Computing*, pages 74–81, 1990.

182. J. Li. *Crystal for Distributed-Memory Machines*. PhD thesis, Yale University, October 1991.

183. J. Li and M. Chen. Compiling communication-efficient programs for massively parallel machines. *IEEE Transactions on Parallel and Distributed Systems*, 2(1):361–376, July 1991.

184. Z. Li, P.H. Mills, and J.H. Reif. Models and resource metrics for parallel and distributed computation. In *Proceedings of 28th Annual Hawaii International Conference on System Sciences, IEEE Press*, 1995.

185. D. J. Mavriplis. Three dimensional unstructured multigrid for the euler equations. In *AIAA 10th Computational Fluid Dynamics Conference*, pages paper 91-1549cp, June 1991.

186. K. McManus. *A Strategy for Mapping Unstructured Mesh Computational Mechanics Programs onto Distributed Memory Parallel Architectures*. PhD thesis, Centre for Numerical Modelling and Process Analysis School of Computing and Mathematical Science, University of Greenwich, London, UK, February 1996.

187. Computing Surface CS-2: Technical overview. TR-S1002-10M115.01A, 1993.

188. J. H. Merlin. ADAPTing Fortran 90 array programs for distributed memory architectures. In *First International Conference of ACPC, Salzburg, Springer-Verlag, LNCS 591*, pages 184-200, September 1991.

189. M. Metcalf and C. Reid. *Fortran 90/95 Explained*. Oxford University Press, 1996.

190. Ethan L. Miller and Randy H. Katz. RAMA: a file system for massively-parallel computers. In *Proceedings of the Twelfth IEEE Symposium on Mass Storage Systems*, pages 163-168, 1993.

191. Jason A. Moore, Philip J. Hatcher, and Michael J. Quinn. Stream*: Fast, flexible, data-parallel I/O. Technical Report 94-80-13, Oregon State University, 1994. Updated September 1995. Appeared at ParCo'95.

192. E. Morel and C. Renvois. Global optimization by suppression of partial redundancies. *Communication of the ACM*, 22:96-103, February 1979.

193. E. Morel and C. Renvois. Interprocedural elimination of partial redundancies. In *S. S. Muchnick and N. D. Jones, editors, Program Flow Analysis: Theory and Applications, Chapter 6*, pages 160-188, 1981.

194. T. C. Mowry. *Tolerating Latency Through Software-Controlled Data Prefetching*. PhD thesis, Standford University, March 1994.

195. S. A. Moyer and V. S. Sunderam. PIOUS: A scalable parallel I/O system for distributed computing environments. In *Proc. Scalable High Performance Computing Conference*, pages 71-78, May 1994.

196. T. A. Mueck. The DiNG - a parallel multiattribute file system for deductive database machines. In *Proc. 3rd Int. Symp. on Database Systems for Advanced Applications*, pages 115-122. World Scientific, April 1993.

197. T. A. Mueck and J. Witzmann. Multikey index support for tuple sets on parallel mass storage systems. In *Proceedings of the 14th IEEE Symposium on Mass Storage Systems*, pages 136-144. IEEE Press, September 1995.

198. *High Performance Fortran Mapper. User's Guide*, October 1995.

199. D. M. Nicol and D. R. O'Hallaron. Improved algorithms for mapping pipelined and parallel computations. *IEEE Transactions on Computers*, 40:295-306, March 1991.

200. Nils Nieuwejaar and David Kotz. The Galley parallel file system. In *Proceedings of the 10th ACM International Conference on Supercomputing*, May 1996. To appear.

201. Nils Nieuwejaar and David Kotz. Low-level interfaces for high-level parallel I/O. In Ravi Jain, John Werth, and James C. Browne, editors, *Input/Output in Parallel and Distributed Computer Systems*, chapter 9, pages 205-223. Kluwer Academic Publishers, 1996.

202. Bill Nitzberg. Performance of the iPSC/860 Concurrent File System. Technical Report RND-92-020, NAS Systems Division, NASA Ames, December 1992.

203. Mark H. Nodine and Jeffrey Scott Vitter. Optimal deterministic sorting in parallel memory hierarchies. Technical Report CS-92-38, Brown University, August 1992.

204. M. G. Norman and P. Thanisch. Models of machines and computation for mapping in multicomputer. *ACM Computing Surveys*, 25:263–30, September 1993.

205. B. Nour-Omid, A. Raefsky, and G. Lyzenga. Solving finite element equations on concurrent computers. In *Proceedings of Symposium on Parallel Computations and their Impact on Mechanics, Boston*, pages 395–403, December 1987.

206. M. Paleczny, K. Kennedy, and C. Koelbel. Compiler support for out-of-core arrays on parallel machines. In *Proceedings of the Seventh Symposium on the Frontiers of Massively Parallel Computation, McLean, VA*, pages 110–118, February 1995.

207. D. M. Pase, T. MacDonald, and A. Meltzer. MPP programming model. Technical report, Cray Research, March 1992.

208. David Patterson, Garth Gibson, and Randy Katz. A case for redundant arrays of inexpensive disks (RAID). In *Proceedings of the ACM SIGMOD International Conference on Management of Data*, pages 109–116, June 1988.

209. David A. Patterson and John L. Hennessy. *Computer Organization & Design. The Hardware/Software Interface*. Morgan Kaufman, 1994.

210. R. H. Patterson, G. A. Gibson, E. Ginting, D. Stodolsky, and J. Zelenka. Informed prefetching and caching. Technical Report CMU-CS-95-134, Carnegie Mellon University, April 1995.

211. *PGHPF. User's Guide*, October 1995.

212. P. Pierce. A concurrent file system for a highly parallel mass storage system. In *Proceedings of the Fourth Conference on Hypercube Concurrent Computers and Applications*, pages 155–160. Golden Gate Enterprises, Los Altos, CA, March 1989.

213. J. S. Plank. *Efficient Checkpointing on MIMD Architectures*. PhD thesis, Princeton University, January 1993.

214. R. Ponnusamy. *Runtime Support and Compilation Methods for Irregular Computations on Distributed Memory Machines*. PhD thesis, Dept. of Computer Science, Syracuse University, Syracuse, NY, May 1994.

215. R. Ponnusamy, J. Saltz, A. Choudhary, Y.-S. Hwang, and G. Fox. Runtime-compilation techniques for data partitioning and communication schedule reuse. Technical Report CS-TR-93-32, University of Maryland, April 1993.

216. Ravi Ponnusamy, Yuan-Shin Hwang, Joël Saltz, Alok Choudhary, and Geoffrey Fox. Supporting irregular distributions in FORTRAN 90D/HPF compilers. Technical Report CS-TR-3268 and UMIACS-TR-94-57, UMD, May 1994.

217. R. Ponnusamy et al. A manual for the CHAOS runtime library. Technical report, University of Maryland, May 1994.

218. J. T. Poole. Preliminary survey of I/O intensive applications. Technical Report CCSF-38, Scalable I/O Initiative, Caltech Concurrent Supercomputing Facilities, Caltech, 1994.

219. T. W. Pratt, J. C. French, P. M. Dickens, and S. A. Janet. A comparison of the architecture and performance of two parallel file systems. In *Proceedings of the Fourth Conference on Hypercube Concurrent Computers and Applications*, pages 161–166. Golden Gate Enterprises, Los Altos, CA, 1989.

220. Apratim Purakayastha, Carla Schlatter Ellis, David Kotz, Nils Nieuwejaar, and Michael Best. Characterizing parallel file-access patterns on a large-scale multiprocessor. In *Proceedings of the Ninth International Parallel Processing Symposium*, pages 165–172, April 1995.

221. D. A. Reed, R. D. Olson, R. A. Aydt, T. M. Madhyastha, T. Birkett D. W. Jensen, B. A. Nazief, , and B. K. Totty. Scalable performance environments for parallel systems. In *Sixth Distributed Memory Computing Conference, Portland, Oregon*, April 1991.

222. A. Rogers and K. Pingali. Process decomposition through locality of reference. Technical Report TR88-935, Cornell University, Ithaca, NY, August 1988.
223. M. Rosenblum and J. K. Ousterhout. The design and implementation of a log-structured file system. *ACM Transactions on Computers*, pages 26–52, February 1992.
224. M. Rosing, R. W. Schnabel, and R. P. Weaver. Expressing complex parallel algorithms in DINO. In *Proceedings of the 4th Conference on Hypercubes, Concurrent Computers*, pages 553–560, 1989.
225. M. Rosing, R. W. Schnabel, and R. P. Weaver. The DINO parallel programming language. Technical Report CU-CS-457-9, University of Colorado, Boulder, April 1990.
226. R. Ruehl and M. Annaratone. Parallelization of Fortran code on distributed-memory parallel processors. In *Proceedings of the 4th International Conference on Supercomputing, Amsterdam*, pages 342–353, 1990.
227. Kenneth Salem and Hector Garcia-Molina. Disk striping. In *Proceedings of the IEEE 1986 Conference on Data Engineering*, pages 336–342, 1986.
228. J. Saltz, K. Crowley, R. Mirchandaney, and H. Berryman. Run-time scheduling and execution of loops on message passing machines. *Journal of Parallel and Distributed Computing*, 8(2):303–312, 1990.
229. K. Sanjari and P. Brezany. Functional specification of the advanced PREPARE parallelization engine. Technical Report Vienna-9018-AdvFuncSpec (PREPARE Project Document), Institute for Software Technology and Parallel Systems, University of Vienna, September 1995.
230. D. Schneider. Application I/O and related issues on the SP/2. *SN Newsletter*, 6, September 1994.
231. K. E. Seamons, Y. Chen, P. Jones, J. Jozwiak, and M. Winslett. Server-directed collective I/O in Panda. In *Proceedings of Supercomputing '95*, December 1995.
232. Elizabeth Shriver and Mark Nodine. An introduction to parallel I/O models and algorithms. In Ravi Jain, John Werth, and James C. Browne, editors, *Input/Output in Parallel and Distributed Computer Systems*, chapter 2, pages 31–68. Kluwer Academic Publishers, 1996.
233. L. M. Silva and J. G. Silva. Global checkpointing for distributed programs. In *Proceedings of the 11th Symposium on Reliable Distributed Systems*, pages 155–162, 1992.
234. L. M. Silva, B. Veer, and J. G. Silva. Checkpointing SPMD applications on transputer networks. In *Proceedings of the Scalable High Peformance Computing Conference, Knoxville, USA*, pages 694–701, 1994.
235. T. P. Singh and A. Choudhary. ADOPT. A Dynamic scheme for Optimal PrefeTching in Parallel File Systems. Technical report, Department of Electrical Engineering and Computer Engineering, Syracuse University, June 1994.
236. M. Snir. Proposal for IO. *Posted to HPFF I/O Forum*, July 1992.
237. L. Snyder. Type architecture, shared memory and the corollary of modest potential. In *Annual Review of Computer Science*, pages 289–317. Annual Reviews Inc., 1986.
238. L. Snyder. Experimental validation of models of parallel computation. In *Computer Science Today, Recent Trends and Developments*, LNCS 1000, pages 78–100. Springer Verlag, 1992.
239. L. Snyder. Foundations of practical programming languages. In *Proceedings of the 2nd ACPC Conference, Gmunden, Austria*, pages 115–134, 1993.
240. NA Software. MDB: NA Software HPF Debugger., October 1995.

241. K. Sridharan et al. Parallel structuring of programs containing I/O statements. In *In Proc. of the 1989 International Conference on Parallel Processing*, volume 2, pages 98–106, 1989.

242. J. Subhlok and T. Gross. Task parallelism in Fx. Technical Report CMU-CS-94-112, School of Computer Science, Carnegie Mellon University, Pittsburgh, PA 15213, 1994.

243. J. Subhlok, J. Stichnoth, D. O'Halloran, and T. Gross. Exploiting task and data parallelism on a multicomputer. In *Proceedings of the 2nd ACM SIGPLAN Symposium on Principles and Practice of Parallel Programming, San Diego, CA*, pages 13–22, May 1993.

244. R. Thakur, R. R. Bordawekar, and A. N. Choudhary. Compiler and runtime support for out-of-core programs. In *Proceedings of the ICS'94*, pages 382–391. ACM, July 1994.

245. R. Thakur, A. Choudhary, R. Bordawekar, S. More, and S. Kudatipidi. Passion: Optimized I/O for parallel systems. *IEEE Computer*, June 1996.

246. Rajeev Thakur and Alok Choudhary. Accessing sections of out-of-core arrays using an extended two-phase method. Technical Report SCCS-685, NPAC, Syracuse University, January 1995.

247. The MPI-IO Committee. MPI-IO: a parallel file I/O interface for MPI, April 1996. Version 0.5.

248. C. W. Tseng. *An Optimizing Fortran D Compiler for MIMD Distributed-Memory Machines*. PhD thesis, Rice University, Houston, January 1993.

249. M. Ujaldon, E. L. Zapata, B. M. Chapman, and H. P. Zima. New data-parallel language features for sparse matrix computations. Technical Report TR 95-2, Institute for Software Technology and Parallel Systems, University of Vienna, January 1995.

250. L. Valiant. A bridging model for parallel computation. *Communications of the ACM*, 33(8):103–111, August 1990.

251. P. J. M. van Laarhoven and E. H. L. Aarts. *Simmulated Annealing: Theory and Applications*. D. Reidel Publishing Company, 1987.

252. A. Veen and M. de Lange. Overview of the PREPARE project. In *4th International Workshop on Compilers for Parallel Computers, Delft*, pages 345–350, December 1993.

253. J. S. Vitter and E. A. M. Shriver. Algorithms for parallel memory I: Two-level memories. *Algorithmica*, 12(2/3):110–147, August and September 1994.

254. J. S. Vitter and E. A. M. Shriver. Algorithms for parallel memory II: Hierarchical multilevel memories. *Algorithmica*, 12(2/3):148–169, August and September 1994.

255. T. von Eicken, D. E. Culler, S. C. Goldstein, and K. E. Schauser. Active messages: A mechanism for integrated communication and computation. In *Proceedings of the 19th Annual International Symposium on Computer Architecture*, pages 256–266, May 1992.

256. R. von Hanxleden. Handling irregular problems with Fortran D. In *Proc. of the Forth Workshop on Compilers for Parallel Computers*, pages 353–364, Netherland, December 1993. Delft.

257. R. von Hanxleden. *Compiler Support for Machine-Independent Parallelization of Irregular Problems*. PhD thesis, Center for Research on Parallel Computation, Rice University, December 1994.

258. R. von Hanxleden and K. Kennedy. A code placement framework and its application to communication generation. Technical Report CRPC-TR93337-S, Center for Research on Parallel Computation, Rice University, Houston, October 1993.

259. R. von Hanxleden, K. Kennedy, C. Koelbel, R. Das, and J. Saltz. Compiler analysis for irregular problems in Fortran D. In *Proceedings of the Third Workshop on Compilers for Parallel Computers*, Vienna, Austria, July 1992.

260. Ko-Yang Wang and D. Gannon. Applying AI techniques to program optimization for parallel computers. In *Book: Parallel Processing for Supercomputers and Artificial Intelligence*, McGraw-Hill, pages 441–485, 1989.

261. M. Weiser. Program slicing. *IEEE Transactions on Softw. Eng.*, 10(4):352–357, July 1984.

262. J. Wiedermann. Quo vadetis, parallel machine models? In *Computer Science Today, Recent Trends and Developments*, Lecture Notes om Computer Science 1000, pages 101–114. Springer Verlag, 1996.

263. A. Wilschut. *Parallel Query Execution in a Main-Memory Database Systems*. PhD thesis, University Twente, Holland, April 1993.

264. R. Wismueller. Debugging of globally optimized programs using data flow analysis. *ACM SIGPLAN Notices. Proc. of the Conference on Programming Language Design and Implementation*, 29(6):278–289, June 1994.

265. M. E. Wolf. *Improving Locality and Parallelism in Nested Loops*. PhD thesis, Standford University, August 1992.

266. M. E. Wolf and M. S. Lam. A data locality optimizing algorithm. *IEEE Transactions on Parallel and Distributed Systems*, July 1991.

267. M. E. Wolf and M. S. Lam. A loop transformation theory and an algorithm to maximize parallelism. *IEEE Transactions on Parallel and Distributed Systems*, October 1991.

268. M. J. Wolfe. More iteration space tiling. In *Proc. Supercomputing '89*, November 1989.

269. J. Wu, J. Saltz, H. Berryman, and S. Hiranandani. Distributed memory compiler design for sparse problems. Technical Report ICASE Report No. 91-13, ICASE, NASA Langley Research Center, Hampton, VA, January 1991.

270. P. T. Zellweger. An Interactive Source-Level Debugger for Control-Flow Optimized Programs. In *ACM Proceedings of the Software Engineering Symposium on High-Level Debugging, SIGPLAN Notices, 18*, pages 159–171, Santa Cruz, CA, August 1983.

271. H. Zima, H. Bast, and M. Gerndt. SUPERB: A tool for semi-automatic MIMD/SIMD parallelization. *Parallel Computing*, 6:1–18, 1988.

272. H. Zima, P. Brezany, and B. Chapman. SUPERB and Vienna Fortran. *Parallel Computing*, 20:1487–1517, 1994.

273. H. Zima, P. Brezany, B. Chapman, P. Mehrotra, and A. Schwald. Vienna Fortran – a language specification version 1.1. Technical Report ACPC-TR 92-4, ACPC Austria, March 1992.

274. H. Zima and B. Chapman. *Supercompilers for Parallel and Vector Computers*. Addison-Wesley, New York, 1990.

275. H. Zima et al. Aurora - advanced models, applications and software systems for high performance computing. Project proposal for the Austrian Special Research Program, July 1996.

276. L. W. Zurawski and R. E. Johnson. Debugging Optimized Code with Expected Behaviour. Unpublished Manuscript, April 1991.

Index

Bold face page numbers are used to indicate pages with important information about the entry, e.g., the precise definition of a term or a detailed explanation, while page numbers in normal type indicate a textual reference.

Lecture Notes in Computer Science

For information about Vols. 1–1143

please contact your bookseller or Springer-Verlag

Vol. 1182: W. Hasan, Optimization of SQL Queries for Parallel Machines. XVIII, 133 pages. 1996.

Vol. 1183: A. Wierse, G.G. Grinstein, U. Lang (Eds.), Database Issues for Data Visualization. Proceedings, 1995. XIV, 219 pages. 1996.

Vol. 1184: J. Waśniewski, J. Dongarra, K. Madsen, D. Olesen (Eds.), Applied Parallel Computing. Proceedings, 1996. XIII, 722 pages. 1996.

Vol. 1185: G. Ventre, J. Domingo-Pascual, A. Danthine (Eds.), Multimedia Telecommunications and Applications. Proceedings, 1996. XII, 267 pages. 1996.

Vol. 1186: F. Afrati, P. Kolaitis (Eds.), Database Theory - ICDT'97. Proceedings, 1997. XIII, 477 pages. 1997.

Vol. 1187: K. Schlechta, Nonmonotonic Logics. IX, 243 pages. 1997. (Subseries LNAI).

Vol. 1188: T. Martin, A.L. Ralescu (Eds.), Fuzzy Logic in Artificial Intelligence. Proceedings, 1995. VIII, 272 pages. 1997. (Subseries LNAI).

Vol. 1189: M. Lomas (Ed.), Security Protocols. Proceedings, 1996. VIII, 203 pages. 1997.

Vol. 1190: S. North (Ed.), Graph Drawing. Proceedings, 1996. XI, 409 pages. 1997.

Vol. 1191: V. Gaede, A. Brodsky, O. Günther, D. Srivastava, V. Vianu, M. Wallace (Eds.), Constraint Databases and Applications. Proceedings, 1996. X, 345 pages. 1996.

Vol. 1192: M. Dam (Ed.), Analysis and Verification of Multiple-Agent Languages. Proceedings, 1996. VIII, 435 pages. 1997.

Vol. 1193: J.P. Müller, M.J. Wooldridge, N.R. Jennings (Eds.), Intelligent Agents III. XV, 401 pages. 1997. (Subseries LNAI).

Vol. 1194: M. Sipper, Evolution of Parallel Cellular Machines. XIII, 199 pages. 1997.

Vol. 1195: R. Trappl, P. Petta (Eds.), Creating Personalities for Synthetic Actors. VII, 251 pages. 1997. (Subseries LNAI).

Vol. 1196: L. Vulkov, J. Waśniewski, P. Yalamov (Eds.), Numerical Analysis and Its Applications. Proceedings, 1996. XIII, 608 pages. 1997.

Vol. 1197: F. d'Amore, P.G. Franciosa, A. Marchetti-Spaccamela (Eds.), Graph-Theoretic Concepts in Computer Science. Proceedings, 1996. XI, 410 pages. 1997.

Vol. 1198: H.S. Nwana, N. Azarmi (Eds.), Software Agents and Soft Computing: Towards Enhancing Machine Intelligence. XIV, 298 pages. 1997. (Subseries LNAI).

Vol. 1199: D.K. Panda, C.B. Stunkel (Eds.), Communication and Architectural Support for Network-Based Parallel Computing. Proceedings, 1997. X, 269 pages. 1997.

Vol. 1200: R. Reischuk, M. Morvan (Eds.), STACS 97. Proceedings, 1997. XIII, 614 pages. 1997.

Vol. 1201: O. Maler (Ed.), Hybrid and Real-Time Systems. Proceedings, 1997. IX, 417 pages. 1997.

Vol. 1202: P. Kandzia, M. Klusch (Eds.), Cooperative Information Agents. Proceedings, 1997. IX, 287 pages. 1997. (Subseries LNAI).

Vol. 1203: G. Bongiovanni, D.P. Bovet, G. Di Battista (Eds.), Algorithms and Complexity. Proceedings, 1997. VIII, 311 pages. 1997.

Vol. 1204: H. Mössenböck (Ed.), Modular Programming Languages. Proceedings, 1997. X, 379 pages. 1997.

Vol. 1205: J. Troccaz, E. Grimson, R. Mösges (Eds.), CVRMed-MRCAS'97. Proceedings, 1997. XIX, 834 pages. 1997.

Vol. 1206: J. Bigün, G. Chollet, G. Borgefors (Eds.), Audio- and Video-based Biometric Person Authentication. Proceedings, 1997. XII, 450 pages. 1997.

Vol. 1207: J. Gallagher (Ed.), Logic Program Synthesis and Transformation. Proceedings, 1996. VII, 325 pages. 1997.

Vol. 1208: S. Ben-David (Ed.), Computational Learning Theory. Proceedings, 1997. VIII, 331 pages. 1997. (Subseries LNAI).

Vol. 1209: L. Cavedon, A. Rao, W. Wobcke (Eds.), Intelligent Agent Systems. Proceedings, 1996. IX, 188 pages. 1997. (Subseries LNAI).

Vol. 1210: P. de Groote, J.R. Hindley (Eds.), Typed Lambda Calculi and Applications. Proceedings, 1997. VIII, 405 pages. 1997.

Vol. 1211: E. Keravnou, C. Garbay, R. Baud, J. Wyatt (Eds.), Artificial Intelligence in Medicine. Proceedings, 1997. XIII, 526 pages. 1997. (Subseries LNAI).

Vol. 1212: J. P. Bowen, M.G. Hinchey, D. Till (Eds.), ZUM '97: The Z Formal Specification Notation. Proceedings, 1997. X, 435 pages. 1997.

Vol. 1213: P. J. Angeline, R. G. Reynolds, J. R. McDonnell, R. Eberhart (Eds.), Evolutionary Programming VI. Proceedings, 1997. X, 457 pages. 1997.

Vol. 1214: M. Bidoit, M. Dauchet (Eds.), TAPSOFT '97: Theory and Practice of Software Development. Proceedings, 1997. XV, 884 pages. 1997.

Vol. 1215: J. M. L. M. Palma, J. Dongarra (Eds.), Vector and Parallel Processing – VECPAR'96. Proceedings, 1996. XI, 471 pages. 1997.

Vol. 1216: J. Dix, L. Moniz Pereira, T.C. Przymusinski (Eds.), Non-Monotonic Extensions of Logic Programming. Proceedings, 1996. XI, 224 pages. 1997. (Subseries LNAI).

Vol. 1217: E. Brinksma (Ed.), Tools and Algorithms for the Construction and Analysis of Systems. Proceedings, 1997. X, 433 pages. 1997.

Vol. 1218: G. Păun, A. Salomaa (Eds.), New Trends in Formal Languages. IX, 465 pages. 1997.

Vol. 1219: K. Rothermel, R. Popescu-Zeletin (Eds.), Mobile Agents. Proceedings, 1997. VIII, 223 pages. 1997.

Vol. 1220: P. Brezany, Input/Output Intensive Massively Parallel Computing. XIV, 288 pages. 1997.

Vol. 1222: J. Vitek, C. Tschudin (Eds.), Mobile Object Systems. Proceedings, 1996. X, 319 pages. 1997.

Vol. 1224: M. van Someren, G. Widmer (Eds.), Machine Learning: ECML-97. Proceedings, 1997. XI, 361 page· 1997. (Subseries LNAI).

Vol. 1226: B. Reusch (Ed.), Computational Intelli· Proceedings, 1997. XIII, 609 pages. 1997.